FV

Schnabel · Polymer Degradation

W. Schnabel

Polymer Degradation

Principles and Practical Applications

Distributed in the United States of America by
Macmillan Publishing Co., Inc., New York

and in Canada by
Collier Macmillan Canada, Ltd., Toronto

Hanser International

Prof. Dr. Wolfram Schnabel
Hahn-Meitner-Institut für Kernforschung Berlin
Bereich Strahlenchemie

Distributed in USA by
Scientific and Technical Books
Macmillan Publishing Co., Inc.
866 Third Avenue, New York, N. Y. 10022

Distributed in Canada by
Collier Macmillan Canada Distribution Center,
539 Collier Macmillan Drive, Cambridge, Ontario

CIP-Kurztitelaufnahme der Deutschen Bibliothek

Schnabel, Wolfram:
Polymer Degradation: Principles and pract.
applications/Wolfram Schnabel. — München;
Wien: Hanser, 1981.
ISBN 3-446-13264-3
ISBN 0-02-949640-3 Macmillan Publishing Co., Inc., New York

Library of Congress Catalog Card Number 81-85158

Cover design: C.A. Loipersberger

Copyright © Carl Hanser Verlag München Wien 1981
Satz und Druck: VEB Druckhaus Köthen
Printed in the German Democratic Republic

This book is dedicated

to my wife Hildegard
and
to my mother Elisabeth

Preface

This book is based on a lecture course on "Degradation and Stabilization of Polymers" given repeatedly during the last few years at the "Technische Universität Berlin". Its objective is, therefore, a concise treatment of the various factors causing chemical changes in polymers resulting in a deterioration of their physical properties. The book aims at elaborating principles of polymer degradation with the intention of giving insight into the mechanism of frequently occurring, generally complicated degradation processes. Quite often, these processes can be considered as "concerted actions" of various modes of physical and chemical attack. Environmental attack on plastics is a typical example. In this case chemical changes are induced by physical processes such as the absorption of sun light or mechanical impact. Although the physical properties are initially only slightly affected, the material then becomes susceptible to further changes, such as in reactions with atmospheric oxygen or enhanced absorption of light. Most prominent, in this respect, is certainly the initiation of autoxidative chain reactions, which usually lead to a rapid deterioration of important physical properties.

Because of the complex nature of many important degradation processes it is rather difficult to subdivide this broad subject. For the purpose of elaborating principles of polymer degradation, an order of subdivision based on important modes of degradation initiation appeared to be appropriate. Accordingly, this book contains chapters on thermal, mechanical, photochemical, radiation chemical, biological and chemical degradation. Each chapter begins with basic facts and subsequent sections depict special aspects pertinent to the mode of initiation under consideration. Various practical applications of degradation are stressed with each mode, such as the synthesis of heat resistant polymers and recycling of scrap plastics in connection with thermal degradation, decomposition of plastics in waste depositories and synthesis of biodegradable polymers in connection with biodegradation, synthesis and application of plastics with adjustable lifetime in connection with photodegradation, the capability of certain polymers to act as drag reducers and stress-induced chemical reactions in connection with mechanical degradation, the use of electron beams and synchrotron radiation in lithographic processes applied to the production of microelectronic devices and to contact microscopy in connection with high energy radiation degradation.

This book would never have been completed without help and support of many people, who are gratefully acknowledged here. The author is indebted to many colleagues for their advice and willingness to supply relevant material. Very valuable information has been obtained from *J. Brandrup, K. Ebert, E. J. Goethals, A. Heuberger, H. Höcker, W. Hofmann, S. J. Huang, W. Kaminsky, R. Kerber, E. Küster, Z. Osawa, G. Schmahl, R. Schulze-Kadelbach, H. R. Schulten, H. Sotobayashi, S. Speth, O. Vogl, K. H. Wallhäuser* and moreover from Bayer AG, Sparte Polyurethane-Forschung and Sparte Kautschuk-Anwendungstechnik, Batelle-Institute (Frankfurt), Verband Kunststofferzeugende Industrie (Frankfurt), and Wirtschaftsverband der deutschen Kautschukindustrie (Frankfurt).

The following persons were involved in the final stages of completion of the manuscript: (i) Mrs. *G. Snoei*, who heroically typed the entire text and never became frustrated with the many corrections and alterations, (ii) Mrs. *H. Gadewoltz*, who skilfully prepared the drawings and (iii) Mrs. *L. Katsikas* and Mr. *R. B. Frings* who, with perseverance, critically read the manuscript.

The author is, moreover, grateful to the management of the Hahn-Meitner-Institut, Berlin, for exempting him from administrative work for an appreciable period.

Last but not least, because many nights and week-ends had to be sacrificed for this book, the patience of the author's family must be praised.

Berlin, Spring 1981   *Wolfram Schnabel*

Contents

1 Introduction

1.1 Definitions

With regard to materials composed of synthetic macromolecules, the term polymer degradation is used to denote changes in physical properties caused by chemical reactions involving bond scission in the backbone of the macromolecule. In linear polymers, these chemical reactions lead to a reduction in molecular weight, i.e. to a diminution of chain length:

$$\sim\!\sim \xrightarrow[\text{scission}]{\text{main-chain}} \sim\!\sim \; + \; \sim\!\sim \qquad\qquad (1.1)$$

When considering biopolymers, the definition of polymer degradation is extended to include changes of physical properties, caused not only by chemical but also by physical reactions, involving the breakdown of higher ordered structures.

In both cases the term polymer degradation involves a deterioration in the functionality of polymeric materials, which in the case of biopolymers is usually called denaturation.

This book is essentially devoted to chemical aspects of polymer degradation. In this connection, it must be pointed out that alterations in physical properties are, of course, not only caused by bond scissions in the polymer backbone, but also by very often simultaneously occurring chemical reactions in pendant groups or side-chains. With the exception of intermolecular crosslinking, however, reactions in pendant groups of linear polymers, affect the physical properties only to a minor extent relative to reactions in the backbone.

Intermolecular crosslinking, i.e. the formation of new chemical bonds between individual macromolecules, may be considered the opposite of degradation as it leads to an increase in molecular size and, at higher conversions, to certain kinds of superstructures *) with characteristic physical properties. In general, intermolecular crosslinking will not be treated in this book except in those cases where degradation and crosslinking occur simultaneously.

Since the splitting of chemical bonds in the backbone or main-chain of linear polymers, i.e. main-chain degradation, is the chief objective of this book, it seems appropriate to point out that the expressions scission, rupture, breakage and lesion are used synonymously to indicate bond fracture. The term lesion is sometimes preferred by biochemists and biologists when referring to bond fracture in biological macromolecules.

Expressions in other languages, equivalent to the term "degradation", as defined above are: dégradation (French), degradación (Spanish), degradazione (Italian), деградация (Russian), Abbau (German), bunkai 分 解 (Japanese).

---

*) spatial networks

1.2   Modes of Polymer Degradation

According to the definition given in Section 1.1 polymer degradation is mainly caused by chemical bond scission reactions in macromolecules. It does not appear meaningful, therefore, to distinguish between different modes of polymer degradation. For practical reasons, however, it is useful to subdivide this broad field according to its various modes of initiation. These comprise thermal, mechanical, photochemical, radiation chemical, biological and chemical degradation of polymeric materials.

Chemical degradation refers, in its strict sense, exclusively to processes which are induced under the influence of chemicals (e.g. acids, bases, solvents, reactive gases etc.) brought into contact with polymers. In many such cases, a significant conversion is observed, however, only at elevated temperatures because the activation energy for these processes is high.

Thermal degradation refers to the case where the polymer, at elevated temperatures, starts to undergo chemical changes without the simultaneous involvement of another compound. Often it is rather difficult to distinguish between thermal and thermo-chemical degradation because polymeric materials are only rarely chemically "pure". Impurities or additives present in the material might react with the polymeric matrix, if the temperature is high enough.

Biologically initiated degradation also is strongly related to chemical degradation as far as microbial attack is concerned. Microorganisms produce a great variety of enzymes which are capable of reacting with natural and synthetic polymers. The enzymatic attack of the polymer is a chemical process which is induced by the microorganisms in order to obtain food (the polymer serves as a carbon source). The microbial attack of polymers occurs over a rather wide range of temperatures. Optimum proliferation temperatures as high as 60 or 70 °C are not uncommon.

Mechanically initiated degradation generally refers to macroscopic effects brought about under the influence of shear forces. Apart from the important role polymer fracture plays in determining the applications of plastics, it should also be pointed out, that stress-induced processes in polymeric materials are frequently accompanied by bond ruptures in the polymer main-chains. This fact can be utilized for example for the mechanochemical initiation of polymerization reactions with the aim of synthesizing block- and graft-copolymers.

Light-induced polymer degradation, or photodegradation, concerns the physical and chemical changes caused by irradiation of polymers with ultraviolet or visible light. In order to be effective, light must be absorbed by the substrate. Thus, the existence of chromophoric (light absor bing) groups in the macromolecules (or in the additives) is a prerequisite for the initiation of photochemical reactions. Generally, photochemically important chromophores absorb in the UV range (i.e. at wave lengths below 400 nm). The importance of photodegradation of polymers derives, therefore, from the fact that the ultraviolet portion of the sunlight spectrum can be absorbed by various polymeric materials. The resulting chemical processes may lead to severe property deteriorations.

High energy radiation such as electromagnetic radiation (X-rays, γ-rays) or particle radiation (α-rays, fast electrons, neutrons, nuclear fission products), is not specific with respect to absorption. The existence of chromophoric groups is not prerequisite as in the case of photodegradation since all parts of the molecule are capable of interacting with the radiation. The extent and character of chemical and physical changes depend on the chemical composition of the irradiated material and on the nature of the radiation.

High energy radiation-induced alterations of polymeric materials are important for their utilization in fields of high radiation flux, e.g. in nuclear reactors. A great body of useful applications is based on the fact that the absorption of high energy radiation causes the generation of reactive intermediates (free readicals and ions) in the substrate. Thus, high energy irradiation is a method quite generally applicable for the initiation of chemical reactions occurring via free radical or ionic mechanisms.

The strong inter-relationship between the various modes of polymer degradation should be emphasized. Frequently, circumstances prevail that permit the simultaneous occurrence of various modes of degradation. Typical examples are: (a) environmental processes, which involve the simultaneous action of UV light, oxygen and harmful atmospheric emissions or (b) oxidative deterioration of thermoplastic polymers during processing, which is based on the simultaneous action of heat, mechanical forces and oxygen.

The aim of this book is to improve the reader's understanding of these rather complex processes by considering the various modes of polymer degradation in separate chapters.

1.3 Mechanistic Aspects

1.3.1 Single Step Processes and Chain Reactions

This section deals with the general features of chemical reactions, occurring during the degradation of macromolecules. As has been shown above, chemical reactions can result from various modes of initiation. Mechanistically, free radical reactions can be distinguished from ionic reactions and from those involving electronically excited states as intermediates. Phenomenologically, however, kinetic differences are more important and, therefore, degradation reactions are classified either as single step reactions or as chain reactions.

In single step reactions, the reaction rate is directly proportional to the rate of initiation. Typical examples are given in Table 1.1. There are photochemical reactions where one main-chain bond is ruptured per absorbed photon, as in the case of ketone polymers (see Section 1.3.3 and Section 4.2). Other examples pertain to enzymatic attack, such as the reaction of amylase with starch. If an amylase molecule interacts with amylose in starch, one glucoside bond is broken per attack (see Chapter 6.3).

The characteristic feature of chain reactions is the self-propagation of the processes, once started. In other words, the initiation reaction yields products that are themselves capable of undergoing spontaneous reactions with intact molecules etc. Under continuous initiation, the reaction rate is accelerated, i.e. the conversion increases exponentially with reaction time. This is indicated in Fig. 1.1, where the oxygen uptake during a photochemical reaction is plotted vs. the time of irradiation. At prolonged reaction times, the autoacceleration is usually followed by an autoretardation stage owing, e.g., to a depletion of the O_2 concentration in the interior of the sample or an inhibition of the propagation by reaction products. The importance of chain reactions derives from

Table 1.1   Typical examples of single step and chain reactions occurring during the degradation of polymers

Type of Reaction	Process	Mode of Initiation
Single Step Reaction	Norrish Type II-reactions in ketone polymers	photochemical
	solvolysis of ester linkages	chemical
	enzymatic attack of peptide and glucoside linkages	biological
Chain Reaction	autoxidation	thermal photochemical mechanical, chemical
	depolymerization of polyolefins	"pure" thermal, at elevated temperature also mechanical, photochemical etc.
	elimination of HCl from polyvinylchloride	thermal

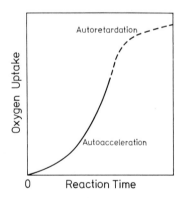

Fig. 1.1   Autoxidation of polymers. Schematic representation of the oxygen uptake as a function of reaction time

the fact that the kinetic chain length, i.e. the number of propagation steps, started by a single initiation reaction, is frequently rather high. This implies a multiplication of deleterious processes, for example, in depolymerizations and oxidative degradations.

As far as the rate of conversion is concerned, single step processes, obviously, are less important than chain reactions. It should be pointed out, however, that sometimes even low conversions can cause pronounced changes of physical properties, especially if main-chain scission or crosslinking processes occur in linear polymers, whose physical properties are very dependent on the average molecular weight. Let us, for instance, consider the following case: hydrocarbon polymers of a homologous series undergo main-chain scission. The fraction of affected repeating units τ is taken as 10^{-4}, corresponding to a conversion of $10^{-2}\%$. As is shown in Table 1.2, the initial degree of polymerization DP_0 is (at this stage of conversion) diminished by 50% when $DP_0 = 10^4$, whereas the corresponding change is as little as 0.1% when $DP_0 = 10$.

Table 1.2   The relative change of the initial degree of polymerization during main-chain scission at constant conversion (according to Eq. (1.9))

τ (a)	α (b)	DP_0 (c)	DP_E (d)	% change of DP_0 (e)
10^{-4}	1	10^4	5×10^3	50
10^{-4}	10^{-1}	10^3	$9{,}09 \times 10^2$	9,1
10^{-4}	10^{-2}	10^2	$9{,}9 \times 10^1$	1
10^{-4}	10^{-3}	10^1	$9{,}99$	0,1

(a) scissions per repeating unit
(b) scissions per initial macromolecule
(c) initial degree of polymerization ($u_{1,0}$ in Eq. (1.9))
(d) final degree of polymerization (u_1 in in Eq. (1.9))
(e) $(DP_0 - DP_E) \, 10^2/DP_0$

1.3.2   Autoxidation

Among the examples presented in Table 1.1 prominence must be given to oxidative chain reactions, or autoxidations, which proceed by free radical mechanisms. Since free radicals are generated in many initiation reactions and because they mostly react readily with molecular oxygen, autoxidation, in polymers is a quite general phenomenon. It is, therefore, appropriate to discuss its mechanism here in some detail. Scheme 1.1 presents the so-called "basic" autoxidation scheme, which has been proposed for the oxidation of hydrocarbon- and other polymers [1].

Since the generation of macroradicals is not the topic of discussion in this section, the initiation reaction (a), given in Scheme 1.1, which pertains to the special case where macromolecules are attacked by low molecular weight free radicals, is not of importance here. It has to be emphasized, however, that hydroperoxide groups are formed in the propagation step. In rigid matrices, i.e. in polymeric solids, successive propagation steps will be likely to proceed in close vicinity of each other or even intramolecularly [2, 3]. In polypropylene, for example, reaction (1.2) describes the process that occurs:

(1.2)

Scheme 1.1   Free radical mechanism of oxidative main-chain scission of linear polymers

R·:   high or low molecular weight free radical, generated by decomposition of the polymer or an additive

PH: macromolecule

$R· + PH$	$→ RH + P·$	initiation	(a)
$P· + O_2$	$→ P-O-O·$		(b)
$P-O-O· + PH$	$→ P-O-O-H + P·$	propagation	(c)
$P· + P·$	$→$ products		(d)
$R· + P·$	$→$ products		(e)
$R· + P-O-O·$	$→$ products	termination	(f)
$P· + P-O-O·$	$→$ products		(g)
$2P-O-O·$	$→ 2P-O· + O_2$		(h)

The decomposition of neighboring hydroperoxide groups in subsequent degradation processes (thermolysis, photolysis) can give rise to specific bimolecular mechanisms (see Section 4.4.1).

With regard to termination reactions, it should be noted that combination reactions involving peroxyl radicals (reactions (f) to (h) in Scheme 1.1) have been found to occur quite effectively in polymer matrices at room temperature. In some cases, peroxide groups may be formed as a consequence of "cage" reactions

$$2PO_2· \rightleftharpoons P-O-O-O-O-P → \boxed{PO· + O_2 + PO·} \begin{array}{l} \longrightarrow POOP \\ \longrightarrow 2PO· \end{array} \qquad (1.3)$$

cage

Oxyl radicals PO· can also play a deleterious role with respect to physical properties, when main-chain rupture (β-scission) occurs:

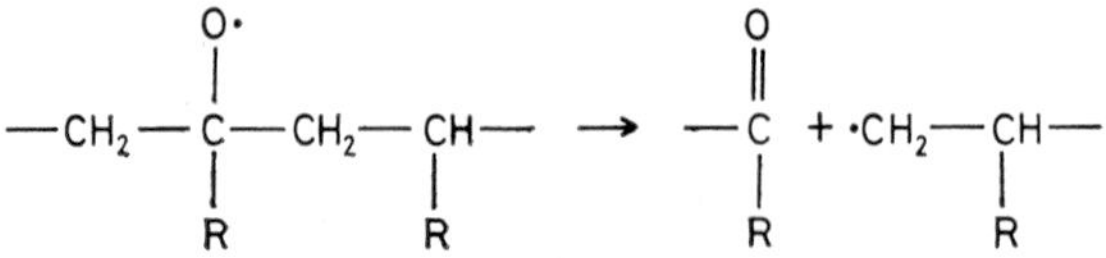

$$-CH_2-\underset{\underset{R}{|}}{\overset{\overset{O·}{|}}{C}}-CH_2-\underset{\underset{R}{|}}{CH}- \quad \rightarrow \quad -\underset{\underset{R}{|}}{\overset{\overset{O}{\|}}{C}} + ·CH_2-\underset{\underset{R}{|}}{CH}- \qquad (1.4a)$$

or generally expressed:

$$PO· → F_1 + F_2· \qquad (1.4b)$$

where F_1 and F_2 denote fragments.

Fragmentation processes, according to (1.4), compete with hydrogen abstraction:

$$PO· + PH → POH + P· \qquad (1.5)$$

According to Geuskens and David [2], in polystyrene, the importance of reaction (1.4) relative to (1.5) increases significantly, as the temperature is increased from 20 to 100 °C indicating that reaction (1.4) is rather unlikely to occur in the rigid matrix.

1.3.3 Random and Specific Site Attack

With degradation reactions of polymers the question arises: Do macromolecules contain sites which are preferentially or exclusively attacked during chemical or physical treatment? Naturally, specific site attack is expected if macromolecules which possess only a single or a few functional groups are brought in contact with a reagent capable of reacting only with these particular functional groups. This type of attack is quite important with polymers containing small portions of impurities incorporated chemically in their pendant groups or backbones.

Very often, the question posed above cannot be answered straightaway. Especially with linear homopolymers, i.e. polymers consisting of chemically identical structural repeating units, it is frequently difficult to figure out whether the probability of a base unit becoming involved in a chemical reaction is equal for all base units in the chain, and, therefore, whether these macromolecules will exhibit random degradation when subjected to a certain chemical or physical treatment.

A typical example of a random process is the photolytical main-chain scission of ketone polymers, such as poly(phenylvinyl ketone), in a Norrish type II process:

$$
\begin{array}{c}
\overset{\displaystyle Ph}{\underset{\displaystyle C=O}{\overset{\displaystyle |}{\underset{\displaystyle |}{}}}} \\
H \qquad C=O \\
-CH_2-\overset{|}{\underset{|}{C}}-CH_2-\overset{|}{\underset{|}{C}}- \quad \xrightarrow{h\nu} \quad -CH_2-\overset{|}{\underset{|}{C}}-H + H_2C=\overset{|}{\underset{|}{C}}- \\
C=O \qquad H \\
Ph
\end{array}
\qquad (1.6)
$$

The repeating unit corresponds to the structure $-CH_2-CH-$. Upon irradiation with

$$-CH_2-\underset{\underset{Ph}{|}}{\overset{\overset{|}{C=O}}{CH}}-$$

light of wavelength $300-370$ nm, the carbonyl groups act as chromophores and, if the polymer sample is thin enough to permit a homogeneous distribution of absorption acts, main-chain scission occurs at random.

Similar situations are encountered in many other cases, where main-chain degradation is induced by the absorption of light, by high energy radiation or via attack by chemical agents.

Non-random main-chain scission has been observed with linear homopolymers subjected to mechanical forces: the center portions of the polymer chains are much more likely to undergo main-chain scissions than other parts of the macromolecules (if certain conditions are met such as high molecular weight and dilute solution). This will be discussed in some detail in Chapter 3.

Moreover, non-random processes must occur with certain block copolymers consisting of long blocks of repeating units A connected by short segments of repeating units B, with the latter being attacked exclusively.

$$A----A-\underbrace{B-B-B-B}-A----A$$
$$\text{sites of attack}$$

2*

It is, furthermore, interesting to note that enzymatic attack seldom proceeds randomly. Normally, the attack is quite specific and, frequently, determined by the conformation of the attacked macromolecule. In this connection, endo- and exo modes of enzyme action, referring to the attack of a linear macromolecule at lateral or terminal sites of the chains, respectively, can be distinguished. Examples will be discussed in Chapter 6.

Problably the most intriguing problems concerning non-random degradation processes refer to so-called "weak links", which are mostly identical with impurities, incorporated chemically in macromolecules. Terminal olefinic unsaturations, contained in many linear vinylpolymers, are a typical example. Quite often the concentration of weak links is so low, that they can hardly be detected by usual analytical methods. However, investigations, of degradation-induced changes of the molecular weight distribution (MWD), have revealed information on the occurrence of non-random processes owing to the presence of weak links. This rather sophisticated method, which has also proved highly valuable for the determination of correct yields for main-chain scissions, will be discussed in the next section.

1.4 Detection of Polymer Degradation

1.4.1 Changes in Molecular Size

Among the various methods for the detection of chemical changes, those based on molecular size determinations play an eminent role, as far as linear (non-crosslinked) polymers are concerned. Table 1.2 shows how in the case of random main-chain scission, even low conversions yield significant changes of the average molecular weight, if the initial molecular weight is sufficiently high. The same holds for intermolecular cross-linking, a process occurring, for example, in many cases where lateral macroradicals are formed:

$$(1.7)$$

$$(1.8)$$

Molecular weight determinations have been carried out, therefore, rather frequently in order to evidence chemical reactions in the main chains or at side groups of linear polymers and to elucidate reaction mechanisms. Since synthetic polymers usually consist of a mixture of chemically identical macromolecules of different size, the molecular size- or molecular weight-distribution (MWD) are important properties which character- ize a polymer. Frequently the MWD is altered upon main-chain rupture and/or cross- linking, hence MWD changes have been treated theoretically by several authors [4—11], especially with respect to random main-chain scission in linear polymers in the absence of crosslinking.

Some of the results pertaining to changes of MWs and MWDs and their value in the determination of scission yields will be presented here. In this connection, it is important to note that MWD determinations can be readily carried out with the aid of gel per- meation chromatography (GPC).

If a linear polymer, having an initial unimodal MWD, undergoes random main-chain scission, the number average degree of polymerization, u_1, decreases with increasing degree of degradation, a, according to Eq. (1.9)

$$\frac{u_1}{u_{1,0}} = \frac{1}{1 + a} \tag{1.9}$$

where a denotes the number of scissions per initial molecule and $u_{1,0}$ the initial number average degree of polymerization. For the change in the weight average degree of polymerization, u_2, as a function of a, Eq. (1.10) holds

$$\frac{u_2}{u_{2,0}} = \frac{2}{a\sigma_0}\left\{1 - \frac{1}{a}\left[1 - \int_0^\infty \frac{\omega(y)}{y/u_{1,0}} e^{-\tau y}\, dy\right]\right\} \tag{1.10}$$

In Eq. (1.10),

$$\sigma_0 = \frac{u_{2,0}}{u_{1,0}}: \text{a parameter characterizing the breadth of unimodal MWDs}$$

$$\omega(y) = \frac{yn(y)}{u_{1,0}}: \text{weight fraction of } y\text{-mers initially present}$$

$$\tau = \frac{a}{u_{1,0}}: \text{number of scissions per base unit}$$

Eq. (1.9) and (1.10) hold for the case of $u_{1,0} \gg 1 \gg \tau$.

MWDs of the Schulz-Zimm [12] type are frequently considered: In this case the weight fraction $\omega(y)$ is expressed by Eq. (1.11)

$$\omega(y) = v^{(b+1)}y^b \frac{e^{-vy}}{\Gamma(b + 1)} \tag{1.11}$$

The parameter v is characterized by the relationships

$$u_1 = \frac{b}{v} \quad \text{and} \quad u_2 = \frac{b + 1}{v} \quad \text{with} \quad b = \frac{1}{\sigma - 1}$$

For a MWD of Schulz-Zimm type, one obtains from Eq. (1.10)

$$\frac{u_2}{u_{2,0}} = \frac{2}{a\sigma_0}\left\{1 + \frac{1}{a}\left[\left(1 + \frac{a}{b_0}\right)^{-b_0} - 1\right]\right\} \tag{1.12}$$

σ depends on a as given by Eq. (1.13) [9]

$$\sigma = \frac{2(1 + a)}{a}\left\{1 - \frac{1}{a}\left[1 - \frac{1}{\left(1 + \frac{a}{b_0}\right)^{b_0}}\right]\right\} \tag{1.13}$$

Criteria for *random main-chain scission* are derived both from Eqs. (1.9) and (1.12), and from Eq. (1.13). Plots representing the relative change $u_2/u_{2,0}$ as a function of the relative change $u_1/u_{1,0}$ at various values of σ_0 are shown in Fig. 1.2. By comparison of the calculated with the experimentally determined molecular weight change it can, therefore, be deduced, whether or not random main-chain scission had occurred.

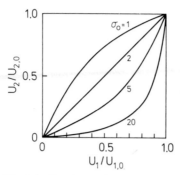

Fig. 1.2 Random scission of polymers with *Schulz-Zimm* type molecular weight distribution at various values of σ_0. Plot of the relative change in weight average molecular weight ($u_2/u_{2,0}$) vs. the relative change in number average molecular weight ($u_1/u_{1,0}$)

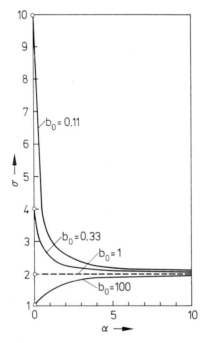

Fig. 1.3 Random scission of polymers with *Schulz-Zimm* type molecular weight distribution. Plot of $\sigma(= u_2/u_1)$ vs. α (number of scissions per initial molecule)

Another criterion for *random scission* is based on the fact, that, independent of σ_0, σ approaches a limiting value of 2 with increasing α. If $b_0 = 1$ *), σ is independent of α

---

*) $b_0 = 1$ corresponds to a so-called most probable distribution

and remains constant during random main-chain scission ($\sigma = \sigma_0 = 2$), as is evident from Eq. (1.13).

Fig. 1.3 shows typical curves demonstrating the dependence of σ on a at various values of σ_0.

With respect to the determination of yields of main-chain scission, it is interesting to note, that for $b_0 = 1$, Eq. (1.12) can be simplified to

$$\frac{u_2}{u_{2,0}} = \frac{1}{1 + a} \qquad\qquad (1.14)$$

1.4.2   Analytical Techniques Applied in Polymer Degradation

Apart from methods for MW- and MWD-determinations, which are powerful in detecting degradation in linear soluble polymers, there is the wealth of conventional analytical methods which are usually also applied in order to demonstrate chemical changes in polymers. An important drawback, in comparison with low molecular weight compounds, is the fact that the separation of reacted from unreacted macromolecules is usually impossible. As most analytical investigations are performed on bulk polymers, which are usually converted by degradation to a very small extent only, this drawback is especially limiting. A somewhat different situation is encountered, however, when gaseous low molecular weight products are formed in the degradation process. These products readily separate from the polymer specimen and can therefore, be easily analyzed (qualitatively and quantitatively). Indeed, volatilization analysis has been frequently applied, especially in thermal, photochemical and radiation-chemical degradations. Sophisticated techniques have been developed to collect, fractionate and analyze volatile products. Systems for volatilization analyses consist normally of a vacuum line, to which a series of cold traps and pressure gauges are connected in order to collect different products. For separation and identification, gas chromatography and mass spectrometry are used.

Commonly, volatilization analysis yields valuable information about chemical reactions which cause decomposition in pendant groups, and this technique contributes, therefore, frequently to the elucidation of degradation mechanisms. The importance of volatilization analysis is, of course, outstanding in cases where the formation of gaseous products is the dominating chemical process, such as in depolymerizations (unzipping) and plasma etching.

As far as chemical changes in bulk polymers are concerned, spectroscopic methods, such as infrared (IR) and ultraviolet (UV) absorption spectroscopy, are used to detect the formation or disappearance of chromophoric groups. Nuclear magnetic resonance (NMR) techniques have proved helpful to analyze structural changes. With regard to the detection of reactive intermediates, prominence must be given to electron spin resonance (ESR) spectroscopy, which allows the detection and usually the identification of free radicals.

There are various other analytical techniques, such as differential thermal analysis (DTA) and differential scanning calorimetry (DSC), which are important in special fields of degradation, for example, DTA and DSC are used in studies of thermal degradation.

It has to be stressed here that this book does not aim at reviewing analytical methods applied in polymer degradation, although occasionally, a certain method will be discussed in some detail. Therefore, for further information, the reader should refer to the literature devoted to polymer analysis [13—19].

On the other hand, it is noteworthy that certain degradation techniques, such as pyrolysis mass spectrometry (see Section 2.7), metathesis and ozonization (see Section 7.3), and solvolysis (see Section 7.2) are of great importance in polymer analysis.

References to Chapter 1

[1] *L. Reich* and *S. S. Stivala*, (a) "Autoxidation of Hydrocarbons and Polyolefins", Dekker, New York (1969);
   (b) "Elements of Polymer Degradation", McGraw-Hill, New York (1971).
[2] *G. Geuskens* and *C. David*, Pure and Appl. Chem. 51, 233 (1979).
[3] *J. C. W. Chien, E. J. Vandenberg,* and *H. H. Jabloner*, J. Polymer. Sci. A-1, 6, 381 (1968).
[4] *E. W. Montroll*, J. Am. Chem. Soc. 63, 1215 (1941).
[5] *E. W. Montroll* and *R. Simha*, J. Chem. Phys. 8, 721 (1940).
[6] *O. Saito*, J. Phys. Soc. Jap. 13, 198, 1451 and 1465 (1958);
   M. Dole (ed.), "The Radiation Chemistry of Macromolecules, Vol. I, Chapter 11, Academic Press, New York (1972).
[7] *V. S. Nanda* and *R. K. Pathria*, Proc. Roy. Soc. (London) A 270, 14 (1962).
[8] *M. Inokuti*, J. Chem. Phys. 38, 1174 (1963).
[9] *M. Inokuti* and *M. Dole*, J. Polymer Sci. A1, 3289 (1963).
[10] *A. Charlesby*, "Atomic Radiation and Polymers", Pergamon Press, Oxford (1960).
[11] *K. W. Scott*, J. Polym. Sci. Symposium 46, 321 (1974).
[12] *L. H. Peebles*, "Molecular Weight Distributions in Polymers", Interscience, New York (1971).
[13] *D. O. Hummel* and *F. Scholl*, "Atlas der Polymer- und Kunststoffanalyse", Hanser, München, 1st ed. Vol. 1—3 (1968); 2nd ed. Vol. 1 (1979).
[14] *B. Ke* (ed.), "Newer Methods of Polymer Characterization", Interscience, New York (1964).
[15] *G. M. Kline* (ed.), "Analytical Chemistry of Polymers", Interscience, New York (1962).
[16] *J. Haslam* and *H. A. Willis*, "Identification and Analysis of Plastics", Van Nostrand, Princeton, N.J. (1965).
[17] *M. Kraft*, "Struktur und Absorptionsspektroskopie der Kunststoffe", Verlag Chemie, Weinheim (1973).
[18] *B. Rånby* and *J. F. Rabek*, "ESR Spectroscopy in Polymer Research", Springer, Berlin (1977).
[19] *M. Hoffmann, H. Krömer* and *R. Kuhn*, "Polymeranalytik I and II", Thieme, Stuttgart (1977).

2 Thermal Degradation

2.1 Introduction

Organic macromolecules as well as low molecular weight organic molecules (consisting essentially of carbon, hydrogen, oxygen and nitrogen) are stable only below a certain limiting temperature range usually from 100° to 200 °C, in special cases a few hundred °C higher. If the temperature is increased to 1 000 °C or higher, organic molecules decompose into small fragments (free radicals, free ions, H_2, CO, etc.).

Polymers consisting of organic macromolecules, therefore, are to be distinguished from many inorganic materials (e.g. certain metals and a variety of silicates) which are resistant up to 2000° or 3000 °C.

The rather high thermal sensitivity of organic substances derives from the fact that molecules are composed of atoms linked together by covalent bonds. The strength of these bonds is limited. Dissociation energies of single bonds in the ground state are in the order of 150—400 kJ per mol at 25 °C (typical values: O—O: 147; C—H: 320—420; C—C: 260—400; C—O: 330 kJ/mol).

At ambient temperature thermal energies correspond to an average value of kT ≈ 2.4 hJ/mol. Thus bond scissions are not feasible at ordinary temperatures, not even at temperatures of a few hundred degrees. One must consider, however, that in condensed systems vibrational energy is rather rapidly dissipated among all molecules and all bonds. If the energy distribution is Maxwellian, a certain fraction of bonds in some molecules will be in a highly excited vibrational state corresponding to energies significantly higher than the average energy. The fraction of highly vibrationally excited bonds increases with increasing temperature. It might then occasionally happen that a repulsive energy level is reached, i.e. that bond breakage occurs.

Usually, the absorption of sufficient energy quanta that exceed the dissociation energy can only occur at temperatures higher than 400 to 600 °C. Thermal cracking of short chain hydrocarbons occurs, at these rather high temperatures and therefore significant degradative conversion of most polymers is also expected. At somewhat lower temperatures (150 to 300 °C), bond scissions are less frequent. However, they can initiate chemical reactions (e.g. oxidations) which proceed more rapidly at these temperatures than at ambient temperature. Frequently, significant conversions can be achieved, therefore, at slightly elevated temperatures, especially if the chemical reactions proceed by a chain mechanism.

Concluding these considerations, it should be emphasized that scissions of chemical bonds under the influence of heat are the result of overcoming bond dissociation energies. While these processes will cause rapid decomposition of polymers only at highly elevated temperatures, the pronounced temperature dependence of the rates of chemical reactions can cause a significant and rather rapid decomposition already under milder conditions. Relevant chemical reactions can be initiated by these bond ruptures, or their occurrence becomes important as activation energies are surpassed, i.e. due to the increase of reaction rates.

Thermal degradation of polymers is an important subject because it covers a broad field, ranging from the development of thermoresistant polymers and ablation problems to the stabilization of thermolabile polymers. Decomposition procedures, such as thermolysis and pyrolysis, have attracted attention and are now utilized for polymer analysis. In view of the expected shortage of raw materials, thermal degradation of polymers is, in several cases, considered an appropriate method to produce certain urgently needed chemicals. As far as recycling of polymer scrap is concerned, thermal degradation is already applied in special cases and might become more attractive in the future as raw material costs for chemicals increase.

The various aspects of thermal degradation will be discussed in the following sections of this chapter.

2.2   Methods for the Evaluation of Heat Resistance

A variety of methods has been developed in order to investigate the thermal stability of plastics. These methods include the measurement of physical properties as well as the detection of chemical changes. In addition, special methods have been introduced in order to measure the rate of degradation as a function of temperature or as a function of time at constant temperature [1–4].

Physical methods concern the transitions (crystallographic transitions, melting, glass transition) which are caused by changes in the morphological structure. These transitions can be detected by measuring the temperature dependence of physical properties such as refractive index, specific volume, specific heat, heat conductance, dielectric constant, etc. Plotting these properties vs. temperature, the transitions are evidenced by more or less pronounced disruptions of otherwise steadily changing values.

The heat distortion of plastics, which is intrinsically correlated to the glass transition temperature of amorphous polymers and to the melting temperature of crystalline polymers, is usually determined by specified procedures. The latter are based on the measurement of a temperature or a temperature range corresponding to a definite distortion of a plastic rod of given size. Procedures frequently referred to include the determination of VICAT temperature [5], MARTENS temperature [6] or heat distortion temperature [7].

Today, thermal analysis [3, 4, 8, 9, 10] is usually carried out with the aid of commercially available equipment allowing the elegant investigation of the thermal behavior of polymers. The most important techniques are differential thermal analysis (DTA) and differential scanning calorimetry (DSC). DTA is a technique in which the temperature difference ΔT between a sample and an inert reference material is recorded as a function of temperature (usually the sample temperature). Typical results are shown in Fig. 2.1 where glass transition and crystalline melting can be recognized. The ΔT values are either positive or negative, depending on whether the transition is exothermic or endothermic.

DSC is a technique which allows the rate of heat transfer (dq/dt) to be recorded as a function of time. The procedure resembles that used for DTA. The ability to determine quantitatively thermal energy flow is useful for a variety of applications (e.g. determinations of the heat of fusion, the specific heat as a function of temperature, the degree of

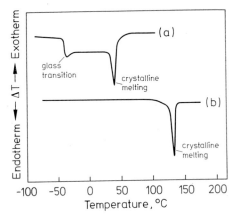

Fig. 2.1   Typical results obtained by differential thermal analysis (DTA) [10].
(a) neoprene W, (b) polyethylene

crystallinity etc.). DSC may also serve as a tool for investigating the oxidative stability of a polymer as shown in Fig. 2.2, where curve 1 was recorded with a polyethylene sample containing no additive, whereas curves 2 and 3 correspond to samples containing different antioxidants. Thus DSC is appropriate for investigating both changes in physical properties as well as chemical changes.

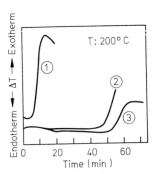

Fig. 2.2   Typical results obtained by differential scanning calorimetry: Polyethylene without (1) and with different antioxidants, (2) and (3). The samples were heated to 200° in the presence of N_2. Subsequent to the addition of O_2 the recording was started [10]

The technical application of plastics at high temperatures requires special tests extending over long periods. Tensile strength and impact strength are usually tested with plastics used in construction. With composites the bending strength is measured, with adhesives the shear strength and with fibers the tenacity, elongation, and the elastic modulus.

The maximum use temperature pertains by definition [11] to a decrease by 10% of the property under investigation (measured at room temperature), when the material is kept continuously at the high temperature (testing period: 8—12 months and extrapolation to 2.5×10^4 hours).

Chemical changes occurring at high temperatures can be detected in special cases by DSC (see above). Of more general importance, however, are the following techniques:

a) Thermogravimetric analysis (TGA) [14], which is based on measuring the temperature dependence of the loss of sample weight due to the formation of volatile products. The loss of weight at constant temperature is frequently measured as a function of time by recording vapor pressure changes in a closed system (isoteniscopy [12]). The onset of degradation is characterized by the temperature at which pressure changes become detectable. Typical examples are shown in Fig. 2.3.

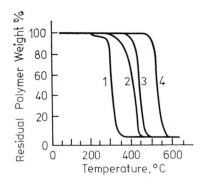

Fig. 2.3 Typical results obtained by thermogravimetric analysis. Plot of the fraction of residual polymer vs. temperature [10].
(1) polyhexafluoropropylene, (2) polypropylene, (3) polyethylene, (4) polytetrafluoroethylene

b) Thermal volatilization analysis (TVA), which is based on the trapping of gaseous products followed by their characterization by the usual methods (UV and IR spectroscopy, elemental analysis etc.) [13]. Mass spectroscopy has been used successfully with TVA by degrading the polymer sample inside the mass spectrometer. Volatile decomposition products are ionized and detected and identified according to their m/e values (mass to charge ratio) [15, 16]. Ionization is achieved either by electron impact (EMS) or by field ionization (FIMS). Typical experimental results will be presented in section 2.7, since TVA can also serve for studying the structure of polymers and the composition of copolymers.

c) Analysis of the nonvolatile residue (determinations of molecular weight and molecular weight distribution, analysis by IR, UV etc.).

2.3 Mechanistic Aspects

The chemical changes occurring during thermal treatment of polymers can be characterized by the following phenomena: chemical bonds in the main chains and in the side chains are ruptured as evidenced by a diminution of the molecular weight and the evolution of low molecular weight gaseous products, respectively. Typical intramolecular reactions are cyclizations and eliminations. In the case of linear polymers, intermolecular crosslinking can occur as indicated by an augmentation of the molecular weight. Linear polymers of the polyolefin type, furthermore, frequently decompose forming large amounts of monomer. This process, which is the more significant the higher the temperature, is denoted as "depolymerization". From the kinetic point of view depolymerization proceeds as a chain reaction (see Chapter 1.3). The strong temperature dependence of the rate of depolymerization is due mainly to an increase in the rate of initiation and also to some extent due to an increase of the kinetic chain length with increasing temperature. Scheme 2.1 depicts in a general form the reactions discussed so far. Apart from depolymerization, other processes, such as radical transfer reactions, can also proceed as chain reactions at elevated temperatures. This has been dealt with in a general way in Chapter 1.3 and will be treated in more detail below.

Scheme 2.1   Schematic depiction of reactions occurring during thermal degradation of polymers

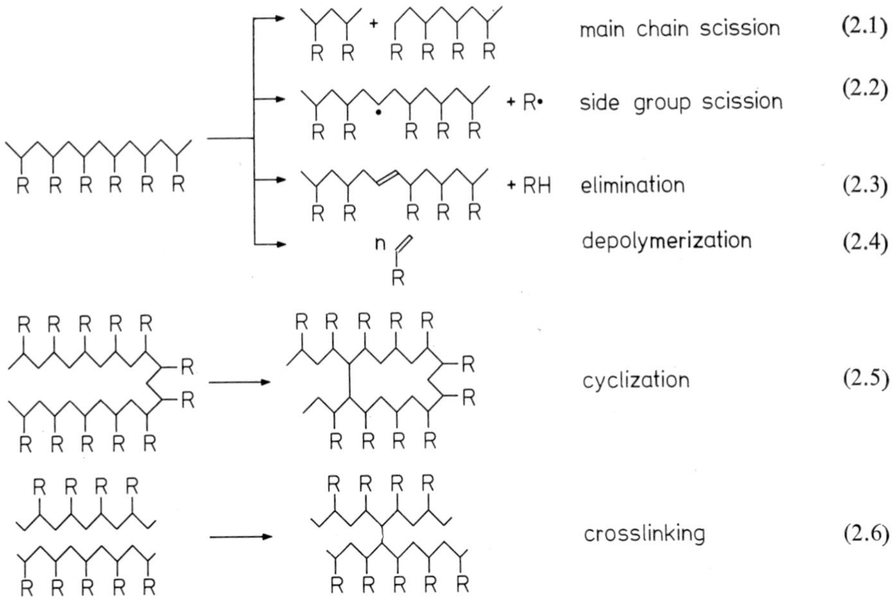

main chain scission   (2.1)

side group scission   (2.2)

elimination   (2.3)

depolymerization   (2.4)

cyclization   (2.5)

crosslinking   (2.6)

With respect to the mechanism of thermal degradation the most important question concerns, of course, *initiation processes*, regardless of whether they are followed up by single step or chain reactions. Scissions of chemical bonds under the influence of heat are rather unspecific in contrast to, e.g., photolytical bond ruptures (see Chapter 4).

As the temperature is increased, the scission probability for all kinds of bonds increases. The bond dissociation energy, however, differs appreciably, as shown in Table 2.1, where selected values are given. Therefore, most polymers contain so-called "weak bonds" which are expected to break with a higher probability than other "stronger" bonds. In linear polymers composed of identical repeating units with equally strong bonds, bond scissions are expected to be distributed statistically along the chains and over all macromolecules in the system. With such polymers, bond breakage, therefore, is a purely stochastic process. Normally, however, the composition of synthetic polymers never corresponds entirely to the molecular formula. The non-ideal polymers we utilize for the production of plastics contain chemically incorporated "impurities", which very often function as weak links. Commercially available polyethylene may contain, e.g., a very small amount of groups of the following structures, where the arrows designate the site of facile bond rupture:

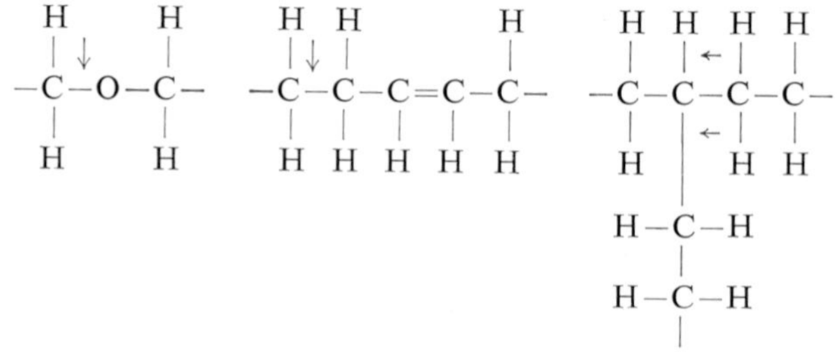

$$
\begin{array}{cccccc}
& H & H & H\ H & H & H\ H\ H\ H \\
& | \downarrow & | & | \downarrow\ | & | & |\ |\ \leftarrow\ |\ | \\
-C & - & O - C - & -C - C - C = C - C - & & -C - C - C - C - \\
& | & | & |\ |\ |\ |\ | & & |\ \ \ \leftarrow |\ \ | \\
& H & H & H\ H\ H\ H\ H & & H\ \ \ H\ H
\end{array}
$$

$$
\begin{array}{c}
H - C - H \\
| \\
H - C - H \\
|
\end{array}
$$

Thermally labile groups are frequently located at the end of linear macromolecules. A typical example are the unsaturated end groups in vinyl polymers

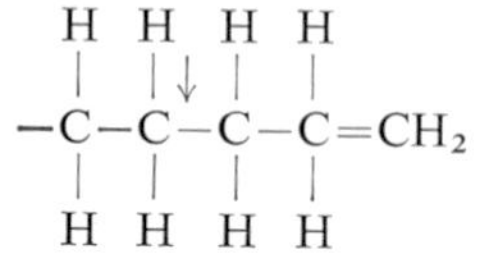

$$
\begin{array}{ccccc}
H & H & H & H \\
| & |\downarrow & | & | \\
-C - & C - & C - & C = CH_2 \\
| & | & | & | \\
H & H & H & H
\end{array}
$$

formed during termination by disproportionation in radical polymerizations. Head-to-head (H—H) linkages, originating from the combination of growing chains, on the other hand, have been also considered as possible weak links. From systematic studies on several homopolymers it is inferred that this seems to be true only in certain cases [69]: H—H and head-to-tail (H—T) polymers of styrene, vinylcyclohexane and methylacrylate exhibit comparable thermal stability as far as onset temperature of degradation and degradation rate are concerned. H—H polymethylcinnamate, however, has a lower onset temperature and a higher degradation rate than the head-to-tail polymer.

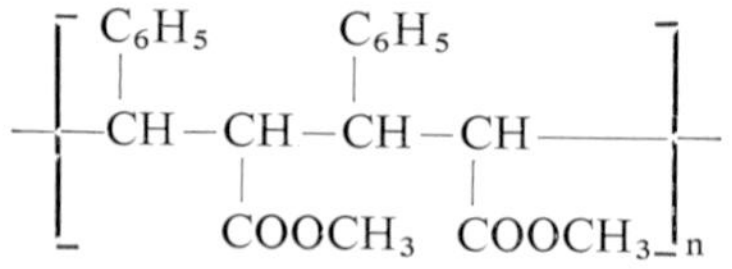

$$
\left[\begin{array}{cc} C_6H_5 & C_6H_5 \\ | & | \\ -CH - CH - CH - CH - \\ | & | \\ COOCH_3 & COOCH_3 \end{array} \right]_n
\qquad
\left[\begin{array}{cc} & C_6H_5 \\ & | \\ -CH - CH - CH - CH- \\ | & |\ \ \ \ | \\ C_6H_5 & O=C\ \ \ C=O \\ & |\ \ \ \ \ | \\ & O\ \ \ \ \ O \\ & |\ \ \ \ \ | \\ & CH_3\ \ CH_3 \end{array} \right]_n
$$

head-to-tail 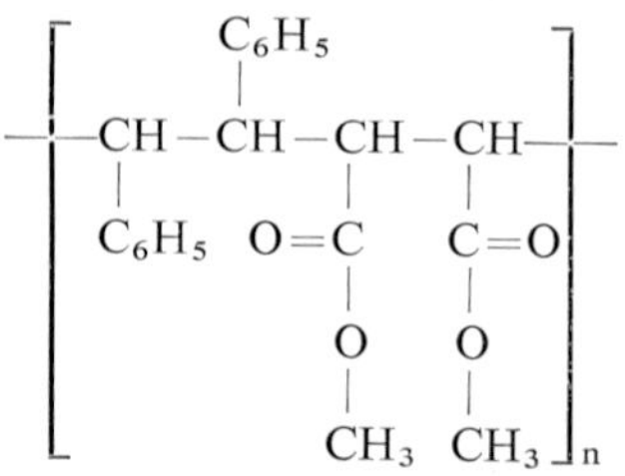   head-to-head

Table 2.1 Selected values of bond dissociation energies E_D
(temperature 25 °C)

bond	E_D (kJ/mol)	reference compound
C=O	729	ketones
C–O	331	$H_5C_2\!\!\downarrow\!\!O-C_2H_5$
C≡C	838	$H-C\!\!\downarrow\!\!\equiv C-H$
C=C	524	$H_2C\!\!\downarrow\!\!=CH_2$
C–C	406	$F_3C\!\!\downarrow\!\!CF_3$
C–C	373	$C_6H_5\!\!\downarrow\!\!CH_3$
C–C	335	$H_3C-\overset{\text{CH}_3}{\underset{\text{CH}_3}{C}}\!\!\downarrow\!\!CH_3$
C–H	507	$H\!\!\downarrow\!\!C\equiv C-H$
C–H	432	$H\!\!\downarrow\!\!CF_3$
C–H	411–427 °	prim. aliphatics
C–H	394	sec. aliphatics
C–H	373	tert. aliphatics
C–H	325	$C_6H_5CH_2\!\!\downarrow\!\!H$

Apart from unimolecular processes, initiation can be due to bimolecular reactions if plastics are composed of two or more substances. At elevated temperatures reactions of the type

$$A + B \rightarrow C \text{ (or } D + E) \tag{2.7}$$

may become possible, the product C (or D and E) being very unstable, i.e. decomposing very rapidly relative to the rate of the reaction $A + B$. The various modes of initiation of thermal degradation processes in linear polymers are compiled in Table 2.2.

In principle, a similar situation is met with thermosetting polymers. However, due to the more or less dense network structure of these polymers the location of ruptured bonds is relatively unimportant. Thus, a differentiation between random and non-random processes becomes rather meaningless. Furthermore, depolymerization (regeneration of starting material) does not occur in this case.

When considerations are restricted to linear polymers being exposed to moderately elevated temperatures, reaction mechanisms generally involve free radicals (species with unpaired electrons), which are usually very unstable and therefore highly reactive.

Table 2.2 Modes of initiation of thermal degradation processes

type of reaction	mode
unimolecular	random main-chain scission and random side-group scission
	rupture of "weak bonds" in the main chain and in side groups
	rupture of bonds at labile groups at chain ends
bimolecular	generation of thermolabile groups or compounds $(A + B \rightarrow C; C \rightarrow R_1' + R_2')$

A general mode of free radical decay corresponds to mutual deactivation, by either recombination or disproportionation

$$P' \cdot + P'' \cdot \rightarrow P' - P'' \quad \text{(recombination)}$$
$$P' \cdot + P'' \cdot \rightarrow P'(H) + P''(-H) \quad \text{(disproportionation)} \quad (2.8)$$

If the system contains compounds with hydrogen atoms, hydrogen abstraction competes with radical-radical deactivation:

$$P' \cdot + P''H \rightarrow P'H + P'' \cdot \quad (2.9)$$

Reaction (2.8) is usually an encounter-controlled process with a low activation energy (determined by the diffusion of reactants). Reaction (2.9) is activation-controlled. At room temperature and at low viscosity of the system, its rate constant is several orders of magnitude smaller than that of reaction (2.8). At ambient temperature reaction (2.8) will be dominant, especially if cage recombination of the macroradicals is possible. Alternatively, at high viscosity, free radicals will be very long-lived (trapped radicals) because the probability for the occurrence of reactions according to (2.8) and (2.9) is extremely low.

At elevated temperatures the situation will frequently reverse. Processes according to (2.9) will occur with high probability after the generation of free radicals during the initiation stage. Since one of the products of reaction (2.9) is a free radical, the whole process is proceeding as a chain reaction. A more complicated mechanism might be encountered if macroradicals undergo unimolecular main-chain rupture before radical transfer reactions take place:

$$\sim\!\!\sim\!\!\overset{\cdot}{\sim}\!\!\sim\!\!\sim \rightarrow \sim\!\!\sim\!\!\cdot \;+\; \sim\!\!\sim \quad (2.10)$$

A sequence of processes according to reactions (2.9) and (2.10) might occur prior to termination according to (2.8) or according to reaction (2.11):

$$P \cdot + X \rightarrow P \text{ (or PH)} + Y \cdot \quad (2.11)$$

$Y \cdot$ denotes a free radical which is not capable of propagating the kinetic chain.

In the presence of molecular oxygen chain reactions are also feasible, the propagation process being characterized by reactions (2.12) and (2.13):

$$P\cdot + O_2 \rightarrow PO_2^{\cdot} \tag{2.12}$$

$$PO_2^{\cdot} + PH \rightarrow PO_2H + P\cdot \tag{2.13}$$

Generally, the free radical reactions discussed so far can be paralleled by depolymerization (reaction 2.4). Depolymerization is characterized by the average zip length z which denotes the average number of monomer molecules formed between initiation and termination. It should be pointed out that the route of the chain reaction may involve several radical transfer and many depolymerization processes (within a single kinetic chain) leading to a diminution of the average degree of polymerization as well as to the evolution of significant amounts of monomer. The kinetics of isothermal depolymerization, with respect to the various parameters (random or chain end initiation, magnitude of z, mode of termination and initial molecular weight distribution) have been treated extensively by other authors [17—23]. A few conclusions are reported here:

With random initiation, the rate of monomer formation decreases linearly with conversion if the average zip length is very large (greater than the average degree of polymerization). Furthermore, the average degree of polymerization of the residual polymer decreases only slightly up to very high degrees of conversion. On the other hand, at low values of z, the rate of monomer formation is only slightly affected by conversion, whereas the degree of polymerization of the residual polymer decreases drastically, even at very low conversions. When initiation occurs at the chain ends, the influence of z on the rate of depolymerization resembles that described above for random initiation, the degree of polymerization is, however, much less drastically affected with increasing conversion. This is illustrated in Fig. 2.4.

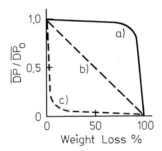

Fig. 2.4   Schematic illustration of the change of the average degree of polymerization DP as a function of the conversion (expressed as weight loss of the polymer) [17—22]. (a) zip length z very high, random and chain end initiation; (b) $z = 0$, chain end initiation; (c) $z = 0$, random initiation

The importance of depolymerization during thermal degradation becomes evident from Table 2.3. PMMA and PαMS are almost quantitatively depolymerized. With other polymers, such as polyisobutene, polystyrene and polybutadiene, depolymerization is an important process, but other types of degradation also occur to an appreciable

Table 2.3   Heat-induced monomer formation of various polymers in the absence of air [23, 24]

Polymer	Monomer Fraction *) (mol %)	Temperature for half-life of 30 min (°C)
Poly-α-methylstyrene (PαMS)	100	
Polyoxymethylene (POM)	100	
Polytetrafluoroethylene (PTFE)	96	510
Polymethylmethacrylate (PMMA)	95	330
Polymethacrylonitrile (PMAN)	85	
Polystyrene (PS)	41	360
Polyisobutene (PIB)	20	350
Polybutadiene (PB)	20	410
Poly(ethylene oxide) (PEO)	4	350
Polymethylacrylate (PMA)	1	330
Polyethylene (PE)	1	
Polyacrylonitrile (PAN)	0	390
Polypropylene (PP)	0	400
Polyvinylchloride (PVC)	0	260
Polyvinylacetate (PVAc)	0	270

*) Fraction of monomer in volatile products

extent. In other cases (polyethylene, polypropylene, polyacrylonitrile, polyvinylchloride) depolymerization does not occur or is detectable only to a negligible extent.

The influence of branching on the thermal stability of *polyolefins* is shown in Fig. 2.5, where thermogravimetric analysis results are presented. It is seen that the stability decreases in the series PE > PP > branched PE > PIB. From the fact that the molecular weight distribution became narrower with increasing conversion, is was inferred that with polyolefins main-chain scission is a random process [27].

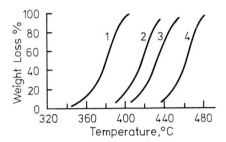

Fig. 2.5   Influence of branching on the thermal stability of polyolefins. Thermogravimetric analysis in the absence of air [26].
(1) polyisobutene, (2) highly branched polyethylene, (3) polypropylene, (4) polyethylene

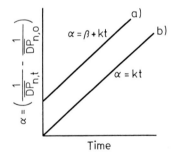

Fig. 2.6   Thermal degradation of polystyrene at $T < 300\,°C$. Schematic illustration of the dependence of the degree of degradation α on the duration of heating. Polystyrene synthesized by free radical polymerization (a) and by ionic polymerization (b). [28–30]

If *polystyrene* (PS) is heated to temperatures above 300 °C volatile products are formed containing monomer (45%) and oligomers. Furthermore, main-chain degradation takes place [2]. Below 300 °C no volatile products are formed. Random main-chain scission occurs as indicated by a linear dependence of the reciprocal degree of polymerization, $(\overline{DP})^{-1}$ on the duration of heat treatment [28–30]. Plots of $(\overline{DP})^{-1}$ vs. time yielded straight lines as shown in Fig. 2.6. However, the plot obtained with radically polymerized PS intersects the ordinate considerably above the origin in contrast to the case of ionically polymerized PS. It was therefore concluded that radically polymerized PS contains weak links in the form of head-to-head structures $-CH_2-CH-CH-CH_2-$

$$\begin{array}{cc} | & | \\ Ph & Ph \end{array}$$

and/or branches which are ruptured very rapidly after the onset of the heat treatment.

Because H—H structures can be discarded as possible weak linkages in this case [69] (see above), branch-points might be solely responsible for the observed effect-provided the radically synthesized polymer did not contain oxygen impurities such as peroxide groups in the backbone ($\sim\sim O-O \sim\sim$),

In the case of *polymethylmethacrylate* the depolymerization is initiated at the chain ends. At temperatures exceeding 270 °C main-chain scissions also contribute to the initiation mechanism [31, 32]. The importance of the zip length on the properties of the residual polymer was clearly demonstrated by thermal treatment of PMMA of different initial degree of polymerization, $\overline{DP_0}$. As can be seen from Fig. 2.7 the degree of polymerization of the residual polymer is independent of the conversion at $\overline{DP_0} = 443$ indicating that $z > DP_0$. At higher values of $\overline{DP_0}$ the degree of polymerization of the residual polymer decreases with increasing conversion which is characteristic for the case $z < \overline{DP_0}$.

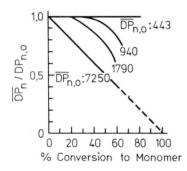

Fig. 2.7   Thermal degradation of polymethylmethacrylate at 220° to 260°C. The number-average degree of polymerization of the residual polymer (expressed as fraction of original $\overline{DP}$) vs. conversion of the polymer to monomer for different initial degrees of polymerization (as indicated in the graph) [32]

The degradation of *polyvinylchloride* is characterized by a very effective elimination of HCl [23, 33, 34]. Depolymerization does not occur. The hydrogen chloride formation commences at a temperature slightly higher than 200 °C if the pyrolysis is carried out in a nitrogen atmosphere. As can be seen from Fig. 2.8, the degradation occurs in two stages as characterized by two plateau regions. The first one is reached at a conversion of about 65 % and pertains essentially to HCl formation. In the second stage the residual polymer containing a high concentration of conjugated C—C double bonds is cross-linked and converted to char. The rate of elimination of HCl depends on various parameters, such as the initial degree of polymerization and the presence of O_2 and additives.

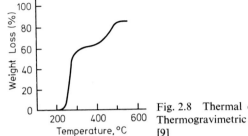

Fig. 2.8   Thermal degradation of polyvinylchloride. Thermogravimetric analysis in the absence of oxygen [9]

The initiation of the elimination process is assumed to be due to the rupture of weak links such as

The elimination is considered to proceed via a free radical chain reaction, the propagation process being:

$$Cl\cdot + -CH_2-CH- \rightarrow HCl + -\overset{\bullet}{CH}-CH- \qquad (2.14)$$
$$||$$
$$ClCl$$

$$-\overset{\bullet}{CH}-CH- \rightarrow -CH=CH- + Cl\cdot \qquad (2.15)$$
$$|$$
$$Cl$$

Furthermore, a molecular elimination mechanism involving HCl as a catalyzing agent appears feasible [35]:

$$\rightarrow 2\,HCl + -CH_2-CH=CH-CH- \qquad (2.16)$$
$$|$$
$$Cl$$

Since PVC is one of the most widely used polymers, stabilization is a crucial problem, to which a tremendous amount of research has been devoted. This aspect will be treated in more detail in Section 2.6.

The few experimental facts described here should be considered as typical examples of the different mechanisms operative during the thermal degradation of polymers. For further reading on the subject several books and articles are recommended [1, 2, 18—23, 36—39].

2.4   Heat Resistant Polymers

A very important aspect of the research concerning the development of modern plastics was, and still is, aimed at synthesizing polymers with a high heat distortion temperature. Metals should be replaced by plastics wherever possible. A great obstacle in approaching this goal turned out to be the fact that the first generation of polymers possessed rather low melting points or glass transition temperatures.

Several routes were followed in order to obtain plastics with increased heat stability: (a) Increase of the degree of crystallinity (in cases where the macromolecules tend to form crystalline regions), (b) Incorporation of polar side groups, (c) Incorporation of aromatic and heteroaromatic rings into the main chain or as side groups, (d) intermolecular chemical crosslinking [40—45].

In principle, such variations were supposed to cause stronger intra- and intermolecular interactions resulting in an increase in glass transition temperatures or melting points. The improvement of the thermal stability implies not only better physical (especially mechanical) properties but also a pronounced resistance towards chemical breakdown (e.g. oxidation) at higher temperatures.

The attainment of chemical resistance at high temperatures was based essentially on the idea of avoiding chemical structures with weak linkages or sites possessing a strong tendency for rearrangement. Further criteria of importance for chemical stability concern resonance stabilisation and structures with unstrained bond angles. Since occasional ruptures of chemical bonds cannot be totally prevented, the breakdown of the chemical structure was expected to be retarded to a significant extent by using double or multiple stranded structures, usually designated "ladder polymers". In a simple case two single strands are chemically linked at regular intervals thus forming a *ladder* polymer, e.g.

With such a polymer, a decomposition process deleterious to the mechanical properties, e.g. decrease of chain length, must involve a number of ruptures per macromolecule significantly exceeding unity (a figure which holds for single-stranded chemical structures).

In other attempts to improve the chemical stability, the chemical structure of polymers was modified by substituting C or H atoms by atoms of other elements such as Si, P, B, F. The most important step in this direction was the discovery of the synthesis of polysiloxanes, having the base unit structure

$$-\underset{\underset{R}{|}}{\overset{\overset{R}{|}}{Si}}-O-$$

Table 2.4 Linear hydrocarbon and fluorinated hydrocarbon polymers

Polymer	Formula of base unit	T_g °C (a)	T_m °C (b)	T_u °C (c)	Processing (d)	Applications
Polyethylene (HD)	$-CH_2-CH_2-$	−110 to −80	110 to 140	90	+	various
Polypropylene	$-CH_2-CH-$ $\quad\quad\vert$ $\quad\quad CH_3$	−15	108 to 212	120	+	various
Polybutene-1	$-CH_2-CH-$ $\quad\quad\vert$ H_3C-CH_2	−45	125 to 140	80	+	pipes, foils, cable insulations, foams
Poly-4-methylpentene-1	$-CH_2-CH-$ $\quad\quad\vert$ $\quad\quad CH_2$ $\quad\quad\vert$ $H_3C-C-CH_3$ $\quad\quad\quad\vert$ $\quad\quad\quad H$	29	230	90	+	transparent packaging materials, sterilizable disposable medical tools
Polyvinylcyclohexane (isotactic)	$-CH_2-CH-$ $\quad H_2\bigcirc H_2$	126	325	220	+	excellent dielectric material $(50-10^{10}$ Hz$)$
Polyvinylfluoride	$-CH_2-CHF-$	155	164 to 235 (e)	140	+	coatings and insulations (buildings)
Polyvinylidenefluoride	$-CH_2-CF_2-$	−46	170	150		insulations, interior coatings of chemical reactors
Polytetrafluoroethylene	$-CF_2-CF_2-$	20	325	200	−	various

(a) glass transition temperature
(b) melting temperature
(c) maximum use temperature
(d) thermoplastic processing possible (+) or impossible (−)
(e) depending on the degree of crystallinity

Table 2.5   Linear carbocyclic aromatic polymers

Polymer	formula of base unit
Poly-p-phenylene	
Poly-p-xylylene	$-CH_2-$⟨O⟩$-CH_2-$
Poly(2,6-di-methyl-1,4-phenylene oxide)	
Poly(sulfone ether)	$-O-$⟨O⟩$-SO_2-$⟨O⟩$-$
Poly(p-phenylene sulfide)	$-S-$⟨O⟩$-$
Polyethyleneterephthalate	$-O-(CH_2)_2-O-\underset{\overset{\|}{O}}{C}-$⟨O⟩$-\underset{\overset{\|}{O}}{C}-$
Poly-p-oxybenzoate	$-O-$⟨O⟩$-\underset{\overset{\|}{O}}{C}-$
Polyamide [1,4- bis (aminomethyl)-cyclohexane + suberic acid]	$-\underset{\overset{\|}{O}}{C}-(CH_2)_6-\underset{\overset{\|}{O}}{C}-NH-CH_2-$⟨H⟩$-CH_2-NH-$
Poly(m-phenylene isophthalamide)	

(a) thermoplastic processing possible (+) or impossible (−)
(b) decomposes at 550°C without melting
(c) can be spun to fibers from organic solutions and pressed at 320°C

T_g °C	T_m °C	T_u °C	Processing (a)	Applications
	(b)	250	—	composites with asbestos or graphite fibers
60—70	405	100	—	coatings of metals, anticorrosive
207	262	200	+	constructional material (electric components)
	288	260	+	coatings
150	287	180 —260	+	coatings of metals, composites with glass fibers for electrical insulations
80	220 —225	130	+	textile fibers
		300	—	self lubricating bearings, electrical insulations
86	296		+	hydrolysis-resistant construction material
270 —280	430	230	— (c)	fibers for fire-resistant fabrics

In the cases of practical importance, R designates methyl or phenyl groups. These polymers, known as "silicones", have maximum use temperatures up to $150-200\,°C$. Linear silicones are, at room temperature, more or less viscous liquids, depending on the molecular weight. Three dimensionally crosslinked silicones are elastomers.

It has to be pointed out that, as the heat distortion temperature of polymers is increased, the processability usually becomes more and more difficult. This holds for the large group of linear polymers with thermoplastic properties. If the melting point is above $200\,°C$ to $300\,°C$, processing is usually accompanied by a rapidly occurring chemical decomposition. Thus, rather stable polymers, though being of linear structure, cannot be processed by the normal methods but have to be treated by special means. In the case of polytetrafluoroethylene and certain other polymers sintering processes are used. Techniques for molding resemble those of powder metallurgy. In other cases (e.g. poly-imides and poly-p-phenylene), the macromolecules are synthesized in situ, i.e. at the site of final use. Molding is not possible in these cases. During the synthesis of heat resistant polymers a compromise between stability and fabricability has always to be found, since the most stable compounds cannot be fabricated or only with great diffi-culty. In any case the fabrication of heat resistant polymers is much more costly than that of thermoplastic and of many thermosetting polymers.

The following examples, discussed briefly, provide insight into the broad spectrum of polymers synthesized for application at high temperatures. The polymers, collected in Tables 2.4 and 2.5, have been selected in most cases for their importance for special applications. None of the high temperature resistant materials is produced on a scale comparable to polyethylene or polypropylene.

It can be seen from Table 2.4 that, among the linear hydrocarbon polymers, only the highly ordered polyvinylcyclohexane possesses a maximum use temperature exceeding $200\,°C$. Substitution of the H atoms in polyethylene by F atoms causes an increase in melting and maximum use temperature. Processing becomes more difficult, however, and polytetrafluoroethylene cannot be processed by thermoplastic methods. Typical examples of linear carbocyclic aromatic polymers have been collected in Table 2.5.
High molecular weight poly-p-phenylene is insoluble and does not melt. Thermal degradation starts at $450\,°C$ in air and at $550\,°C$ in the absence of oxygen. Processing is accomplished by sintering at $400\,°C$ and $14\,000$ bar. Low molecular weight poly-p-phenylene is used for composites with fibers of graphite or asbestos.

In order to facilitate processing polymers of the type

were synthesized, where X denotes a small group or an atom, the insertion of which provide bonds in the chain about which rotation can easily occur. Furthermore, X is not coplanar with the chain axis and, therefore, causes "kinks" in the chain. A great variety of such compounds exist at present and some of them are compiled in Table 2.5. In commercially available products, X corresponds to: $-CONH-$, $-COO-$, $-S-$, $-CH_2-CH_2-$, $-O-$, $-SO_2-$. As far as the onset of gas evolution is concerned, the thermal stability of all of these compounds is inferior to that of poly-p-phenylene. The chemical nature of the linkages between the aromatic rings also exerts a strong influence on the stability as can be seen from TGA results shown in Fig. 2.9 [40]. Some aromatic polyethers such

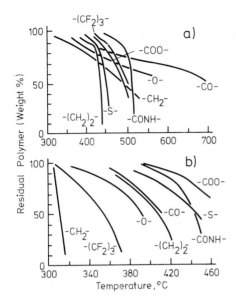

Fig. 2.9   Thermal degradation of phenylene polymers. Weight fraction of residual polymer measured after a 2 hour heat treatment at the temperature given on the abscissa [40].
(a) inert atmosphere, (b) air

as poly-2,6-dimethyl-1,4-phenyleneoxide have become important as construction materials for permanent use at temperatures between 120 and 200°C.

Concerning thermo-oxidative stability, aromatic poly(sulfone ethers) can be considered as being among the most stable thermoplastic materials. In connection with the aromatic rings, the SO_2 group is integrated in a conjugated system having a high degree of resonance stabilization and permitting a rather easy distribution of absorbed external energy. Results of TGA analyses are presented in Fig. 2.10.

Several aromatic polyesters of moderate thermal stability are commercially available, the most important being polyethyleneterephthalate, which is produced on a large scale and is mainly used for the production of textile fibers. Poly(p-oxybenzoate) is of higher thermal stability, but processing is only possible by sintering or other inconvenient techniques.

Many polyamides have been synthesized involving cyclic aliphatic as well as cyclic aromatic amides. To a certain extent, some of them have become commercially important. Examples are given in Table 2.5. These polymers are interesting for their resistance against hydrolysis at moderately elevated temperatures. Polyamide made from 1,4-bis-(aminomethyl)cyclohexane and suberic acid is much more stable towards humidity at elevated temperatures than polyamide-6 or polyamide-6,6. It can, therefore conveniently be used in washing machines etc. Out of the great number of purely aromatic polyamides known, only poly(m-phenylene isophthalamide) is used on a commercial basis. It is utilized as fiber material for fire resistant fabrics and high temperature resistant gas filters etc. Furthermore, high modulus fibers are produced from carbocyclic aromatic

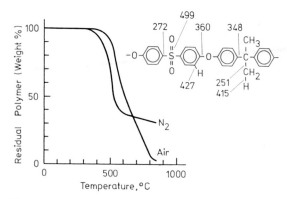

Fig. 2.10 Thermal degradation in the absence and presence of O₂ of a poly(sulfone ether). Structure shown in the graph, numbers denote the dissociation energy (in kJ/mol) [46]

copolyamides, so-called ordered copolymers, with structures such as [47]:

A vast number of polymer structures has emerged from the synthesis of macromolecules containing heterocyclic groups. Table 2.6 shows the heterocyclic structures which have been incorporated into long chains, the heterocyclic rings functioning as repeating units. Also copolymers with alternating heterocycles and p-phenylene rings have been synthesized. The stability of these compounds is determined by the extent of resonance interaction and conjugation between heterocycles and carbocycles. The highest degree of stability is exhibited by polythiadiazole, poly(1,3,4-oxadiazole) and poly-4-phenyl-1,2,4-triazole, with maximum use temperatures of about 300 °C. Polyhydantoine is used on a commercial basis as electrical insulation material in the form of foils and coatings.

Polymers consisting of benzoheterocyclic compounds of the structure

or

Ar: aromatic group
W, X, Y, Z: heteroatoms
(O, S, −N= or −NH−)

are of higher thermal stability than polymers consisting purely of heterocyclic or aromatic moieties. Decomposition starts in an inert atmosphere at 400 to 700 °C. A classification according to the stability in air (TGA at 370 °C) yields the following series: polyimide > polybenzoxazole > polyquinoxaline > polybenzthiazole > poly-N-phenylbenzimidazole > polybenzimidazole.

Table 2.6 Structural repeating units of heterocyclic polymers [37]

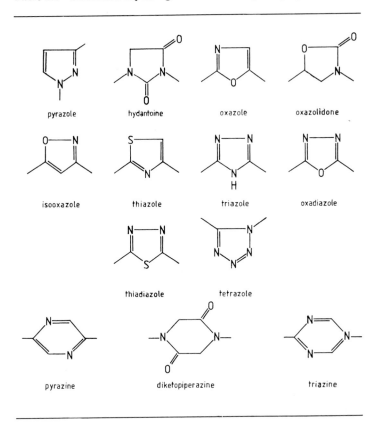

With respect to commercially available polymers, aromatic polyimides range among the most prominent heat resistant polymers. They are synthesized by a multistep process, the last step of which occurs simultaneously with processing to foils, fibers, foams etc. Maximum use temperatures are about 260 °C (short time usage up to 480 °C is possible). Most prominent representatives of this group of polymers are those based on pyromellitic anhydride and aromatic diamines, e.g.

with X: $-S-$, $-SO_2-$, $-CH_2-$, $-CO-$, $-SO-$, $-O-$.

Extremely high heat resistance is exhibited by several "ladder polymers". Typical examples of structural repeating units of these polymers are:

polybenzimidazopyrro-
lone (pyrrone)

polyimidazobenzophenan-
throline

In the absence of oxygen, the decomposition temperatures of these polymers are in the range of 600 °C. However, the oxidative stability is poorer than or only comparable to that of linear polymers of analogous chemical structure with base units connected by single bond linkages. Increased thermo-oxidative stability might emerge from hydrogen-free structures as in the case of the following polyimide [48]:

Ladder structures are also obtained by heat-induced intramolecular cyclization of poly-acrylonitrile:

$$(2.17)$$

If polyacrylonitrile fibers are pyrolized initially at 1 000 °C and subsequently heat-treated at ca. 2 700 °C, so-called graphite fibers of a rather orderly lattice structure are obtained. The carbon fibers are subject to oxidative attack if kept at elevated temperature in the presence of O_2. They do not melt but sublime at 3 650 °C in an inert atmosphere. The density is lower but the thermal conductivity is higher than in the case of silica fibers.

The still developing field of high temperature resistant polymers could not be treated extensively in this section. The reader, therefore, is referred to the literature cited above [40—48]. For detailed information concerning this subject the recently published monograph of *Bühler* [37] is highly recommended.

2.5 Ablation

The penetration of human beings into the space surrounding the earth was accompanied by enormous technological developments in various fields including that of polymers. With respect to protection against the very high temperatures spacecraft, missiles and satellites have to endure during ascent or descent through the atmospheres of the earth or other planets, a new technology has emerged: ablator technology. At the present time polymers or polymer containing composites are the most widely used ablator materials [49–51].

The term ablation here denotes the dissipation (by means of a heat shield which is thermally decomposable) of heat generated by atmospheric friction during re-entry or ascent of a spacecraft. Ablation usually implies loss of matter from the exposed surface. Mass loss occurs in combination with melting, pyrolysis (evaporation), combustion or mechanical erosion.

Ablators are not only used for heat shields in order to protect space crafts, but also for less obvious applications such as shielding of components in rockets, e.g. nozzles. The heat protecting action of ablators becomes comprehensible upon considering the following facts: (a) Heat is very ineffectively conducted by the heat shield. Therefore, the temperature of the ablator rises, (b) Heat is consumed due to physical or chemical changes, e.g. melting and sublimation, or depolymerization and vaporization (decomposition into gaseous compounds), (c) The heated ablator looses energy by emission of radiation (radiant energy loss), (d) Gaseous decomposition products are injected into the boundary layer causing a reduction of the rate of heat transfer to the surface of the ablator.

Polymers were found to be appropriate ablators, possessing low thermal conductivity, low density, high specific heat and thermal decomposability yielding low molecular weight gaseous products.

Ablators are frequently tested with the aid of plasma arcs which create a high temperature ionized gas stream to which the ablator is exposed for a period corresponding to the flight heating time. Details concerning plasmachemical degradation processes will be dealt with in Chapter 5. Space crafts or missiles enter the atmosphere of planets with velocities ranging from about 5 to 60 km s^{-1}, corresponding to huge kinetic energies of about 15 to 180 MJ/kg of vehicle mass. These energies are sufficient to decompose the whole craft into small molecules. Most of the energy is, however, dissipated in the gas layer surrounding the spacecraft.

Most ablators are composite materials. The polymer matrix contains, e.g., fiber reinforcement, low density fillers, gas formers, endothermally easily decomposing compounds etc. Examples of systems used in aeronautics are listed in Table 2.7, which shows that the polymeric matrix consists, in these cases, of a thermosetting material (phenol-formaldehyde or epoxy resins) or a crosslinked elastomeric material (silicones). All of them possess a threedimensional network structure and decompose only at temperatures exceeding 300 °C. Systems containing only a single material are rarely used. Among polymers only polytetrafluoroethylene belongs to this class.

Whereas the characteristic feature of the decomposition of polytetrafluoroethylene is depolymerization, char formation is the most characteristic decomposition phenomenon with the materials compiled in Table 2.7.

Table 2.7 Typical ablators [49]

Material	Polymeric matrix	Fiber	Low density filler	Low temperature Sublimer
Silica-Phenolic	Phenolic	Silica		
Carbon-Phenolic	Phenolic	Carbon		
Nylon-Phenolic	Phenolic	Nylon Fabric		Nylon Fabric
Corkboard	Phenolic		Granular Cork	
Filled silicone (ESA-3560)	Silicone	Glass Fibers	Phenolic Microballoons	
Filled silicone (SLA-220)	Silicone	Silica Fibers	Silica Microspheres	
Thermolag T-230	Phenolic Epoxy			$Mo(CO)_6$

The char layer formed at the surface of the heat shield is rigid and porous. Often it is non-uniform and contains crack patterns due to shrinkage during formation. It provides good heat insulation and thus decreases heat flow into the interior. Therefore, the occurrence of endothermic processes such as vaporization and sublimation is favored underneath the surface. The porous structure of the char layer allows the pyrolysis gas to flow. As a result the gas is heated rapidly and can easily be decomposed. The char layer provides, furthermore, a high temperature outer surface for re-radiation. Char-forming ablators undergo mechanical erosion (spallation). Material is ejected from the surface as particles of various size, bundles or laminae. Sometimes the complete char layer is ejected. Char loss diminishes the ablative properties of the heat shield since it increases the heat flow into the interior and prevents optimum heat conversion.

Since this highly interesting field is treated here only briefly, the reader is referred to an excellent article by *Strauss* [49] which is very valuable as most of the relevant information can otherwise be gathered only from rarely available reports.

2.6 Stabilization

It has been pointed out in the foregoing sections of this chapter that thermal degradation is a prominent obstacle for the application of plastics at high temperatures. In order to overcome this problem, tremendous efforts have been devoted to the synthesis of new heat resistant polymers (see Section 2.4). The successful development of this field was somewhat hampered, however, by the fact that the fabrication of heat resistant polymers is usually very difficult. Therefore, the stabilization of polymers has always received great attention from the producer as well as from the user. Naturally, this interest concerned primarily the most widely applied polymers such as polyolefins and polyvinylchloride.

It has to be realized, of course, that, as the deterioration of physical properties of macromolecules is due to the breakage of chemical bonds, stabilization is not possible, since there are no means of protecting a molecule in a system from receiving energy via mutual collisions. Thus, the situation here differs from other cases, e.g., from that of photodegradation (see Chapter 4), where it is possible to prevent the absorption of energy by certain molecules (screening) or to stimulate energy transfer between molecules of different nature (sensitization).

Since bond breakage by thermolysis, i.e. initiation, cannot be prevented, in principle, the action of a stabilizer is limited to the inhibition or retardation of subsequent processes which frequently proceed as chain reactions. Thus, most stabilizers against thermal degradation normally function as chain terminators, i.e. they are capable of interfering with the propagation reaction, whereas they are uneffective towards the initiation process [36, 39, 52, 53]. Since most frequently heat damage is caused by oxidation processes occurring by a free radical mechanism, many stabilizers belong to the radical scavenger type and are denoted as "antioxidants". Generally speaking, a stabilizer XH acts according to reaction (2.18):

$$P-O-O\cdot + XH \rightarrow P-O-OH + X\cdot \tag{2.18}$$

where X· denotes a rather unreactive free radical which is not capable of propagating the chain reaction. Phenols and amines are generally used as chain terminators by virtue of their ability to donate hydrogen atoms in competition with reaction (2.13)

$$PO_2^\cdot + PH \rightarrow PO_2H + P\cdot \tag{2.13}$$

Table 2.8   Typical stabilizers against thermal oxidation *)

mode of stabilization	compounds
chain terminators	
hydroperoxide decomposers	$(C_{12}H_{25}S)_2$

*) for a list of stabilizers see Ref. [54]

4 Schnabel, Polymer Degr.

Typical stabilizers are presented in Table 2.8 and a possible mechanism illustrating the mode of action of a so-called "hindered phenol" is described by the following reactions:

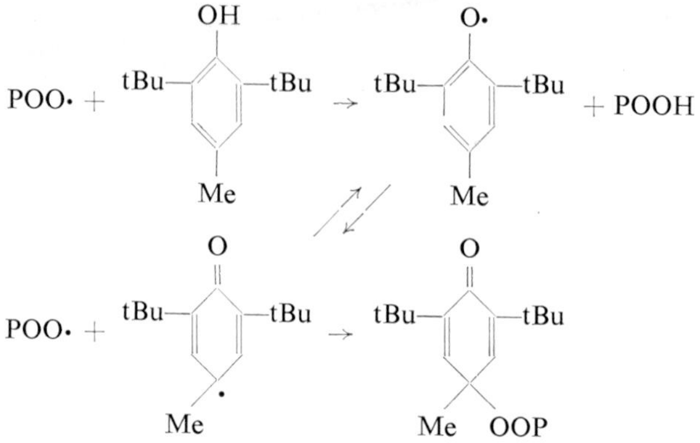

$$
\text{POO}\cdot \; + \quad \underset{\text{Me}}{\overset{\text{OH}}{\underset{\text{tBu}\text{—}\quad\text{—tBu}}{\bigcirc}}} \quad \rightarrow \quad \underset{\text{Me}}{\overset{\text{O}\cdot}{\underset{\text{tBu}\text{—}\quad\text{—tBu}}{\bigcirc}}} \; + \text{POOH} \tag{2.19}
$$

$$
\text{POO}\cdot \; + \quad \underset{\text{Me}}{\overset{\overset{\text{O}}{\|}}{\underset{\text{tBu}\text{—}\quad\text{—tBu}}{\bigcirc}}}^{\cdot} \quad \rightarrow \quad \underset{\text{Me}\quad\text{OOP}}{\overset{\overset{\text{O}}{\|}}{\underset{\text{tBu}\text{—}\quad\text{—tBu}}{\bigcirc}}} \tag{2.20}
$$

Another mode of stabilization concerns the removal of hydroperoxides, which are formed in the propagation step of oxidative chain reactions (2.12) and (2.13). The decomposition of hydroperoxides causes chain branching

$$
\text{P—O—OH} \rightarrow \text{P—O}\cdot + \cdot\text{OH} \tag{2.21}
$$

$$
\text{PO}\cdot + \text{PH} \rightarrow \text{POH} + \text{P}\cdot \tag{2.22}
$$

$$
\cdot\text{OH} + \text{PH} \rightarrow \text{H}_2\text{O} + \text{P}\cdot \tag{2.23}
$$

The products of reaction (2.21) are capable of attacking intact molecules (reactions 2.22 and 2.23) thus generating P· radicals which can react according to reaction (2.12) and hence new oxidative chains are started.

A number of compounds react with peroxides and hydroperoxides rather effectively, especially at elevated temperatures, the most prominent being sulphur-containing compounds, e.g. disulfides.

In the case of diphenyldisulfide it was found that the reaction with hydroperoxides yields thiosulphinates which also react with POOH [53]:

$$
\text{C}_6\text{H}_5\text{SSC}_6\text{H}_5 \xrightarrow{\text{POOH}} \text{C}_6\text{H}_5\overset{\overset{\text{O}}{\|}}{-}\text{S—S—C}_6\text{H}_5 \tag{2.24}
$$

$$
\text{POOH} + \text{C}_6\text{H}_5\overset{\overset{\text{O}}{\|}}{-}\text{S—S—C}_6\text{H}_5 \rightarrow \text{inert products} \tag{2.25}
$$

Synergistic effects were often found when a peroxide decomposer was applied together with a chain terminator. This effect is simply explained by assuming that hydroperoxides formed during the chain propagation, reaction (2.13) are readily converted to inert products in the presence of a hydroperoxide decomposer. In the absence of the latter, reactions (2.21 to 2.23) still occur causing the consumption of additional chain terminator molecules.

A rather specific mode of stabilzations pertains to the protection of polyvinylchloride against the elimination of HCl and the simultaneously occurring discoloration [24, 33,

52—56]. In this case it seems that effective stabilization involves the binding of HCl according to

$$(R-COO)_2Cd + 2HCl \rightarrow 2R-COOH + CdCl_2 \tag{2.26}$$

Frequently used are hexanoates or octanoates of Zn, Cd, Pb, Ba. The absorption of HCl prevents the catalytic action of HCl according to reaction (2.16) and in addition the formation of metal chlorides via the reaction of HCl with metal parts of the surroundings. Certain metal chlorides are capable of catalyzing the decomposition of polyvinylchloride, (see Chapter 7). Since ester formation according to reaction (2.27) parallels HCl absorption, e.g.:

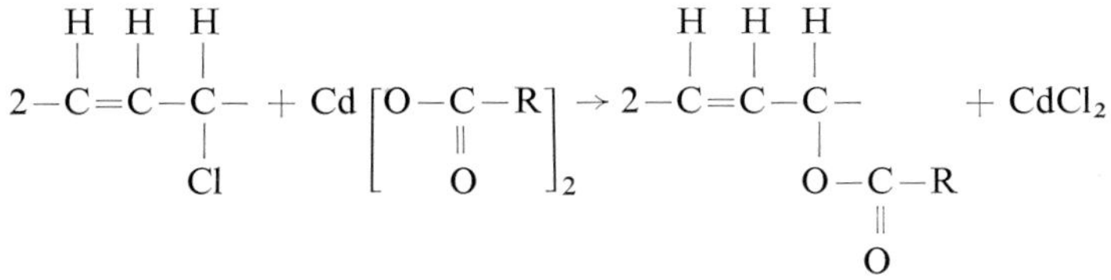

$$2-\overset{\overset{\displaystyle H}{|}}{C}=\overset{\overset{\displaystyle H}{|}}{C}-\underset{\underset{\displaystyle Cl}{|}}{\overset{\overset{\displaystyle H}{|}}{C}}- \ + \ Cd\left[O-\overset{\overset{\displaystyle \|}{C}}{\underset{\displaystyle O}{}}-R\right]_2 \rightarrow 2-\overset{\overset{\displaystyle H}{|}}{C}=\overset{\overset{\displaystyle H}{|}}{C}-\underset{\underset{\displaystyle O-C-R}{|}}{\overset{\overset{\displaystyle H}{|}}{C}}- \ + \ CdCl_2 \tag{2.27}$$

an additional mode of protection emerges [57]. Namely, the ester-exchange reactions occur between labile chlorine atoms in the polymer and the salt. This results in stabilization, since such labile chlorine atoms may be assumed to initiate dehydrochlorination chain reactions upon heating.

Organotin compounds of the general formula

$$(C_4H_9)_2Sn\left(\overset{\overset{\displaystyle O}{\|}}{OC}-R\right)_2, \text{ denoted below as } Bu_2SnY_2,$$

are also used as stabilizers for polyvinylchloride. The structure of the alkyl group R varies widely, e.g. compounds derived from maleic and lauric acid are utilized. It is assumed that the stabilizing action is due to exchange reactions with labile Cl atoms:

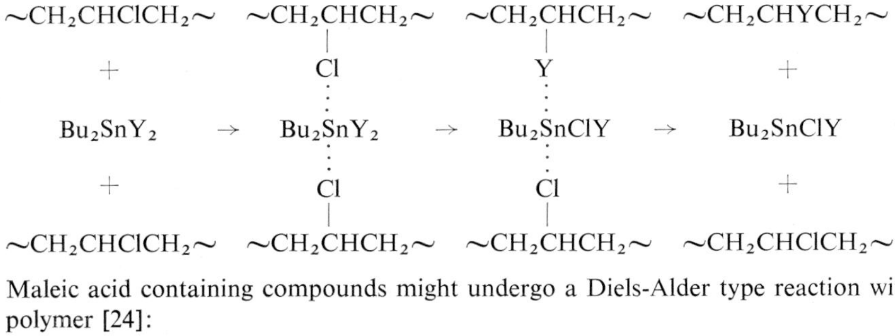

~CH₂CHClCH₂~   ~CH₂CHCH₂~   ~CH₂CHCH₂~   ~CH₂CHYCH₂~
      + | | +
                   Cl   Y
Bu₂SnY₂   →   Bu₂SnY₂   →   Bu₂SnClY   →   Bu₂SnClY   (2.28)
      +   Cl   Cl   +
~CH₂CHClCH₂~   ~CH₂CHCH₂~   ~CH₂CHCH₂~   ~CH₂CHClCH₂~

Maleic acid containing compounds might undergo a Diels-Alder type reaction with the polymer [24]:

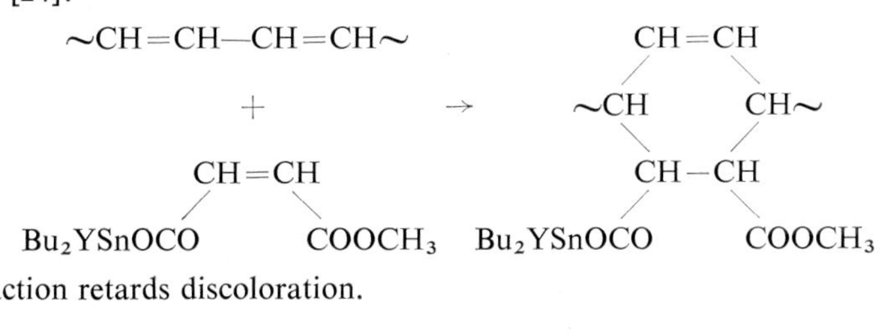

~CH=CH—CH=CH~   CH=CH
                                    /   \
        +   →   ~CH   CH~   (2.29)
                                    \   /
     CH=CH   CH—CH
     /   \
Bu₂YSnOCO   COOCH₃   Bu₂YSnOCO   COOCH₃

This reaction retards discoloration.

4*

Generally, polymer stabilization is not restricted to stabilization against heat but also against other modes of initiation, especially against light of the visible and near UV range. The absorption of light frequently induces free radical chain reactions, propagating the same way as in the case of thermal initiation. With respect to the capability of a substance to act as a chain terminator, it can be used as a heat stabilizer and as a photo-stabilizer. Nevertheless it has to be pointed out that there are photostabilizers which specifically interfere with photoinitiation as is explained in detail in Section 4.4.3.

2.7   Thermal Decomposition and Polymer Analysis

In Section 2.2 mass spectroscopy was mentioned in connection with thermal volatiliz-ation analysis of polymers. It was pointed out that with the aid of several mass spectro-metric techniques it is possible to conveniently identify gas products evolved during thermal decomposition. Fig. 2.11 shows, as a typical example, the mass spectrum of volatile compounds generated during pyrolysis of polystyrene. It exhibits among other peaks those of monomer, dimer and trimer.

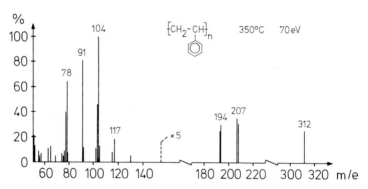

Fig. 2.11   Pyrolysis mass spectrum of polystyrene obtained at $T = 350\,°C$; electron energy: 70 eV [15]

Although studies of this kind have been concentrating on the thermal behavior of polymers only recently [15, 16, 58, 59] several applications of pyrolysis mass spectro-metry promise to be of importance for the analysis of polymers.

Concerning the chemical structure of polymers it should be possible to gain informations, e.g., on head-to-head structures in linear macromolecules. In the case of polystyrene, the existence of head-to-head structures is indicated by the detection of stilbene ions $(Ph-CH=CH-Ph^+)$ in the pyrolysis mass spectrum. An interesting additional problem which has been tackled by pyrolysis mass spectrometry concerns the sequence of bases in nucleic acids, e.g., deoxyribonucleic acids [60].

Furthermore, it is possible to rapidly gain knowledge concerning the composition and the arrangement of sequences in *copolymers*. In the case of a copolyamide of amino-benzoic acid and γ-aminobutyric acid the alternating structure could be demonstrated with the aid of pyrolysis mass spectrometry [61]. The results, which might be also considered as typical examples, are presented in Fig. 2.12. It should be pointed out that calibration with standards is always necessary, which appears to be a limiting factor concerning the general application of the method.

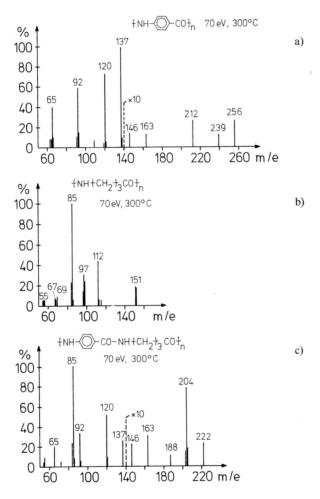

Fig. 2.12   Pyrolysis mass spectra of (a) poly(iminocarbonyl-1,4-phenylene), (b) poly-iminocarbonyltrimethylene, (c) alternating copolyamide. $T = 300\,°C$. Electron energy: 70 eV [61 (b)]

Finally, pyrolysis mass spectrometry might be used to analyze unknown polymer samples, especially if the available amount of material is very small, an important consideration in the field of biomacromolecules.

2.8   Thermal Degradation and Recycling

The fate of plastics scrap and waste has attracted attention recently not only for ecological reasons (dealt with in Section 6.6) but to an increasing extent also from the economic point of view. The fact that crude oil is becoming rare and therefore more costly has stimulated activities around the world concentrating on the question of whether it is possible to utilize plastics scrap and waste as a raw material or an energy source. The recycling of plastics (in its broader sense) has been studied and various methods for industrial scale recycling have been developed. It should be pointed out that the term recycling in its strict meaning refers to the reuse of plastic material which has been used before or which accumulates as scrap during the fabrication of plastic articles. In its broader meaning recycling includes any kind of utilization thus excluding final trans-ference of plastics to waste depositories (rubbish dumps). In Germany (Federal Republic), the fraction of plastics contained in garbage amounts to about 5% [62], a figure which is probably characteristic for many other countries. However, the importance of this figure becomes evident from the fact that, in 1974 the total amount of plastics waste corresponded to about one million tons (metric), of which only about 5%, or 50,000 t, was recycled [62]. Serious obstacles exist which prevent a rapid increase of this recycled fraction. The most serious problem concerns the segregation of plastics waste, since only a small fraction of total garbage consists of plastics. Another important problem pertains to the incompatibility of most polymers. Blends of two or more polymers usually do not form homogeneous systems. Blending usually causes a pronounced diminution of mechanical properties relative to those of articles consisting of only one polymer. This implies that collected plastic articles have to be separated according to the chemical nature of the polymers. Polyethylene articles have to be separated from polystyrene or polyvinylchloride items etc. The separation of plastic articles of different chemical origin by flotation (i.e. sink-swim separation based on differing densities) has been proposed. Manual separation is possible and has been applied. All of these proce-dures are rather costly and, therefore, not economically favorable at the present time. Furthermore, the greatest portion of plastics waste originates in private homes, where the plastics could be separated before disposal. Such procedures appear feasible only if the situation changes drastically (much higher raw material prices, increased sense of civic duty among people, etc.).

Presently, recycling appears economical only in those cases where plastics scrap of identical chemical composition is easily collected, such as during the fabrication of certain articles. Several rather promising procedures will now be described. It should also be pointed out that there exists a rather important possibility for the utilization of plastics in cases where greater amounts of plastics waste can be collected but separation according to the chemical composition is difficult or impossible. This concerns the utiliza-tion of plastics as an energy source. As can bee seen from Table 2.9, the calorific value of several polymers is comparable to that of heating oil. Burning of plastics waste can, therefore, be rather effective, and can be used to develop the high temperatures necessary to operate waste pyrolysis plants.

Other possibilities of utilizing blends of plastics in waste pertain to pyrolysis procedures whereby the polymers are decomposed into a series of low molecular weight compounds which are then separated in subsequent steps. Most important from the economic point

Table 2.9 Calorific values of several materials [64]

Material	Calorific Value (kJ/kg)
Polystyrene	4.6×10^4
Polyethylene	4.6×10^4
Polyvinylchloride	1.9×10^4
Heating Oil	4.4×10^4

of view are at present such processes which are based on the thermal decomposition of chemically homogeneous plastics.

Concerning the most common plastics, only polymethylmethacrylate (PMMA) depolymerizes upon heating with a high yield (see Section 2.3), thus forming the starting material (methylmethacrylate) for the synthesis of PMMA. In other cases the monomer yield is rather low and a mixture of decomposition products is formed. According to a process of Ruhrchemie AG [65], polyethylene is pyrolyzed at about 400 °C resulting in low boiling hydrocarbons*) containing a significant portion of olefins together with waxes and carbon black. In this case polyethylene scrap is admixed to molten hydrocarbon wax kept at the pyrolysis temperature. A flow scheme of the plant is depicted in Fig. 2.13. In a similar way technical processes for the pyrolysis of polyolefins at

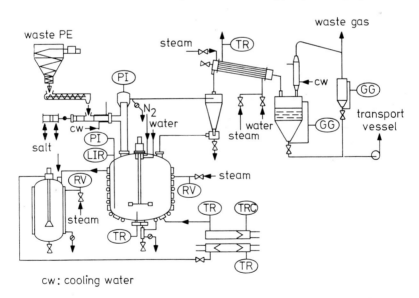

cw: cooling water

Fig. 2.13   Flow scheme of a process for the pyrolysis of polyolefins and other polymers [65].
(TR) temperature recorder, (TRC) temperature recorder and controller, (PI) pressure indicator, (GG) gauge glass, (RV) relief valve, (LIR) level indicator and recorder

---

*) (about 95 %)

temperatures between 300 and 500 °C have been developed by Japanese companies.
There is a process for the production of heating oil (Nichimen) and another for the
production of gasoline of high octane content (Agency Ind. Sci. Techn.). In the latter
case the pyrolysis is performed in an atmosphere of H_2 in the presence of a catalyst.
Oxidized waxes are obtained according to a process developed by Hoechst which proceeds at temperatures between 120° and 140 °C in the presence of O_2. These and several
other processes for the pyrolysis of polypropylene, polystyrene and polyvinylchloride
were compiled in Table 2.10. Pyrolysis at higher temperatures (600—800 °C) is used in
other processes which have been developed, in certain cases, to the pilot plant stage.

Table 2.10   Technical processes for the pyrolysis of various polymers.
(According to *J. Brandrup* [62])

Polymer	Products	Specifications	Producer
Polyolefins	heating oil	300—450 °C Al_2O_3—SiO_2	Nichimen Japan
	95 % liquid products	350—450 °C waxes	Ruhrchemie Germany
	gasoline of high isooctane constant	350—500 °C H_2/catalyst	Agency Ind. Sci. Techn. Japan
	oxidized waxes	120—140 °C O_2	Hoechst Germany
Polypropylene	isobutene	400—650 °C Al-silicates	CHO Rands Japan
Polyvinylchloride	dichloroethane carbontetrachloride	200 °C, Cu(I) 200 °C, Cl_2	Asahi, Japan
	aromatic compounds	350 °C phosphoric acid Na-silicate	Kizo Kogyo, Japan
Polystyrene	styrene	300 °C	Shimada, Japan Yamaguchi, Japan

Molten salt baths or fluidized beds using sand as heat transfer agent are used in these
processes. The pyrolysis products contain a significant fraction of aromatic compounds
even if polyethylene, polypropylene or polyvinylchloride are decomposed. An interesting application of the fluidized bed method is the decomposition of used automobile
tires, a process which has been developed at the University of Hamburg [63, 64] and
which is operated at present in a pilot plant. Table 2.11 shows a compilation of the

pyrolysis products obtained during decomposition of tires and some polymers. It can be seen that the fraction of aromatic compounds is rather high. The gaseous products can be used for heating the reactor thus permitting plant operation without energy supply from the outside. As the particle size does not influence the process, it is assumed that used tires can be pyrolyzed in a technical plant without requiring previous disintegration.

Table 2.11   Pyrolysis products (in weight %) of the decomposition at 740 °C of polyethylene (PE), polystyrene (PS), polyvinylchloride (PVC) and used automobile tires in a fluidized sand bed reactor [64]

Material / Product	PE	PS	PVC	Used Tires
Hydrogen	0.5	0.03	0.7	0.8
Methane	16.2	0.3	2.8	10.2
Ethylene	25.5	0.5	2.1	2.6
Ethane	5.4	0.04	0.4	1.2
Propene	9.4	0.02	0.4	0.7
Isobutene	1.1	—	—	0.2
1,3-Butadiene	2.8	—	—	0.3
Pentene and Hexene	2.0	0.01	—	0.1
Benzene	12.2	2.1	3.5	4.2
Toluene	3.6	4.5	1.1	3.8
Xylene and Ethylbenzene	1.1	1.0	0.2	1.9
Styrene	1.1	71.0	—	2.3
Naphthalene	0.3	0.8	3.1	0.9
Carbon	0.9	0.3	8.8	42.8
Hydrogen Chloride	—	—	56.3	—
Hydrogen Sulfide	—	—	—	1.9
Filler	—	—	—	7.9
Others *)	17.3	15.0	19.3	17.0

*) aromatic and aliphatic compounds of higher molecular weight

Several other processes aimed at the utilization of plastics scrap as raw materials are based on their chemical decomposition at elevated temperature, a subject which is dealt with in more detail in chapter 7. Two such processes are used commercially:

(a) the hydrolysis or methanolysis of polyethyleneterephthalate which yields dimethyl-terephthalate (Hoechst) [62] and (b) the hydrolysis of polyurethanes yielding polyether-glycols and diamines (e.g. tolylenediamine). This process is used by several companies (Bayer, Ford, General Motors, Upjohn) [62]. It should be pointed out that in both of these cases starting materials for the synthesis of the decomposed polymers are obtained.

2.9   Heat Effects in Biopolymers

The effects treated so far in this chapter referred to chemical and physical changes in synthetic polymeric material. The attention of the reader was principally directed to irreversible chemical changes resulting from bond scissions, i.e. from the destruction or variation of the primary structure of macromolecules. It has been pointed out, furthermore, that heat-induced physical changes due to a diminution of the extent of intra- and intermolecular physical interaction can lead to a breakdown of important mechanical properties. However, these changes are in general reversible and are utilized, e.g., for the processing of thermoplastic polymers.

In the case of biopolymers the situation is different. The pronounced temperature sensitivity of most biological systems, due, in most cases, to the fact that heat-induced variations of intra- and intermolecular interactions are irreversible, is well-known. In the case of biomacromolecules these interactions are responsible for the establishment of so-called secondary and tertiary structures. The term "secondary structure" pertains to the conformation or shape of an individual macromolecule which depends ultimately on the allowable rotations in the main-chain and on the interaction of side groups with each other and/or with solvent molecules. The term "tertiary structure" refers to the packing arrangement of the macromolecules, i.e. the formation of superstructures which is based on specific intermolecular interactions. The key to an understanding of the heat sensitivity of biomacromolecules is the fact, that their tertiary structures are closely related to their functions and that slight increases in temperature induce alterations of these superstructures. Since heating-up results in a loss of the biological function of native material, the process of thermal deactivation is designated as "denaturation" by biochemists and biologists. The following are characteristic examples for the denaturation of biomacromolecules.

Lysozyme is a globular protein with a specific three-dimensional structure containing 129 base units (molecular weight ≈ 14.300). It acts as an enzyme in the cleavage of certain polysaccharide chains. Lysozyme contains several short α-helical sections and is intramolecularly crosslinked by four sulfur bonds ($S-S$ bridges) [66]. In its native state the polypeptide chain is folded due to Van der Waals and salt-like interactions as well as some $NH \cdots O=C$ hydrogen bonds, resulting in a deep cleft within the molecule.

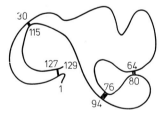

Fig. 2.14   Schematic illustration of the unfolded lysozyme chain. The bars indicate $S-S$ linkages [66]

This cleft contains the active site. The intramolecular interactions are overcome when the temperature is increased and the polypeptide chain then unfolds. Fig. 2.14 shows schematically the unfolded chain which is, however, kept in a rather compact shape due to the S−S bridges. The denaturation can be followed, as indicated in Fig. 2.15 (a) and (b), by measuring the increase of the hydrodynamic radius and the increase of the optical absorption at 274 nm with increasing temperature [66].

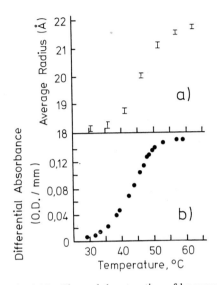

Fig. 2.15   Thermal denaturation of lysozyme [66].
(a) the average hydrodynamic radius and (b) the differential absorbance vs. the temperature. [Lysozyme]: 10 g/l; pH: 1.45 (0.2 M KCl)

Collagen is the most common protein of connective tissue and forms a major part of the matrix of bones. Extractable collagen (tropocollagen) possesses in the native state a three-stranded helical structure, or triple helix. One of the three chains has an amino acid composition somewhat different from that of the other two. Each single chain has a molecular weight of about 3×10^5. The chains are connected by hydrogen bonds formed between NH and CO groups of different strands. At temperatures above 40 °C the tropocollagen molecule dissociates into three chains of equal molecular weight, a process known as the collagen/gelatin transition [67].

Deoxyribonucleic acid (DNA) provides the genetic information required for the function and reproduction of biological organisms. Its primary structure can be schematically depicted by the repeating unit:

$$\left[\begin{array}{c} \text{(2-deoxy-D-ribose)-(phosphate)} \\ | \\ \text{base} \end{array} \right]_n$$

The base is either adenine (A), guanine (G), cytosine (C) or thymine (T). Native DNA possesses a double-strand helix structure in the solid state as well as in solution. As shown in Fig. 2.16 (a) the strands are connected by hydrogen bonds. The double helix is formed via base pairing of complementary DNA strands, as depicted in Fig. 2.16 (b) (adenine always opposite to thymine and guanine to cytosine). Heating of native DNA

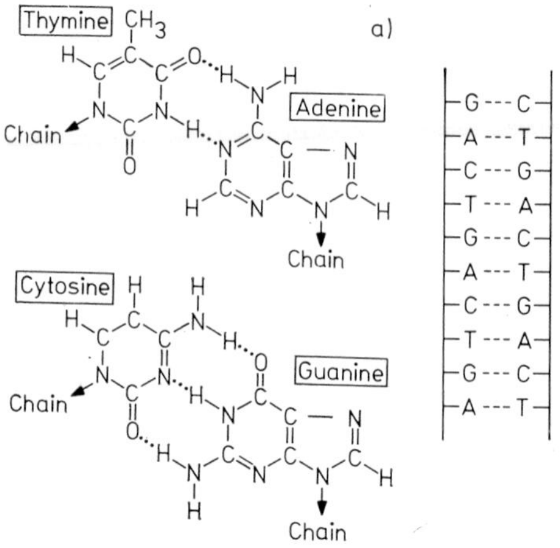

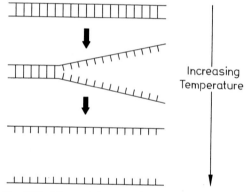

Fig. 2.16   Base pairing in deoxyribonucleic acid (DNA).
(a) hydrogen bonds between thymine and adenine and between cytosine and guanine,
(b) schematic illustration of a double strand DNA molecule

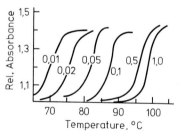

Fig. 2.17   Thermal denaturation of DNA. Diagram illustrates the melting of hydrogen bonds

Fig. 2.18   Thermal denaturation of E. coli K 12-DNA. Relative absorbance at 260 nm vs. temperature at various concentrations of KCl (as indicated in the graph in mol/l) [68]

in solution causes a helix-coil transition: The double helices dissociate into single strand molecules as shown in Fig. 2.17 due to the "melting" of hydrogen bonds and of van der Waals interactions. This process can be studied by following the change of the hypochromicity by measuring the optical absorption at 260 nm as a function of temperature. Typical plots are shown in Fig. 2.18. The temperature corresponding to the inflection point of the curves is known as the melting temperature T_m. It can be seen from Fig. 2.18 that T_m increases with increasing ionic strength [68]. The denaturation of DNA is also achieved at room temperature by changes in pH (acidic or alkaline denaturation).

These three examples, which stand for numerous others, demonstrate the high sensitivity of biomacromolecules towards heat. They show that heating can become deleterious for the biological function at temperatures far below those where the primary structure is attacked, i.e. where chemical reactions are initiated.

References to Chapter 2

[1] *N. Grassie*, "Chemistry of High Polymer Degradation Processes", Butterworth, London (1956).
[2] *S. L. Madorsky*, "Thermal Degradation of Organic Polymers", Polym. Rev. 7, Interscience, New York (1964).
[3] *D. Schultze*, "Differentialthermoanalyse", Verlag Chemie, Weinheim (1969).
[4] *W. J. Smothers* and *Y. Chiang* (ed.), "Differential Thermal Analysis", Chemical Publ. Co., New York (1958).
[5] TGL 17274, DIN 53460, VDE 0302/III.43, ASTMD 1525-85T.
[6] TGL 14071, DIN 53458, DIN 53462.
[7] ASTM-D 648-56, DIN 53461.
[8] *C. B. Murphy*, Anal. Chem. 50, 143 R (1978).
[9] *K. Derge* and *R. Schneider*, Chem. Zeit. 94, 703 (1970).
[10] *P. F. Levy*, DuPont Instruments Reprint, RL-32, Jan. 1970.
[11] DIN 46432, Blatt 2.
[12] *C. Doyle*, WADD Techn. Report 60-283 (1960).
[13] *I. C. McNeill*, Eur. Polym. J. 3, 409 (1967) and 6, 373 (1970).
[14] *N. Grassie*, "M. T. P. International review of science", Butterworth, London (1972).
[15] *I. Lüderwald*, Proc. 5th Europ. Symp. Polym. Spectroscopy, Cologne 1978, *D. O. Hummel* (ed.), Verlag Chemie, Weinheim-New York (1979), p. 217.
[16] *D. O. Hummel*, *H.-D. Schüddemage* and *K. Rübenacker*, "Mass Spectroscopy" in *D. O. Hummel* (ed.), "Polymer Spectroscopy", Verlag Chemie, Weinheim (1974), Chapter 5.
[17] *H. H. G. Jellinek*, "Degradation of Vinyl Polymers", Academic Press, New York (1955).
[18] *H. H. G. Jellinek*, in "Encyclopedia of Polymer Science", Vol. 4, p. 740, Interscience, New York (1966).
[19] *L. A. Wall* in *G. M. Kline* (ed.), "Analytical Chemistry of Polymers-II", Interscience, New York (1962).
[20] *R. H. Boyd*, in *R. T. Conley* (ed.), "Thermal Stability of Polymers", Dekker, New York (1970).
[21] *D. R. Hill* in *S. H. Pinner* (ed.), "Weathering and Degradation of Plastics", Columbine Press, Manchester and London (1966).
[22] *L. R. Reich* and *S. Stivala*, "Elements of Polymer Degradation", McGraw-Hill, New York (1971).
[23] *C. David*, in *C. H. Bamford* and *C. F. H. Tipper* (ed.), "Chemical Kinetiks", Vol. 14, "Degradation of Polymers", Elsevier, Amsterdam (1975).

[24] *L. D. Loan* and *F. H. Winslow*, in *W. L. Hawkins* (ed.), "Polymer Stabilization", Wiley-Interscience, New York (1972).
[25] *G. G. Achhammer, M. Tryon* and *G. M. Kline*, Kunstst.-Plast. 49, 600 (1959).
[26] *F. H. Winslow* and *W. L. Hawkins*, in *R. A. V. Raff* and *K. W. Doak* (eds.), "Crystalline Olefine Polymers", Interscience, New York (1965).
[27] *T. E. Davis, R. L. Tobias* and *E. B. Peterli*, J. Polym. Sci. 56, 485 (1962).
[28] *G. G. Cameron* and *G. P. Kerr*, Eur. Polym. J. 4, 709 (1968) and 6, 423 (1970).
[29] *J. Boon* and *G. Challa*, Makromol. Chem. 84, 25 (1965).
[30] *A. Nakajima, F. Hamada* and *T. Shimizu*, Makromol. Chem. 90, 229 (1966).
[31] *I. C. McNeill*, Eur. Polym. J. 4, 21 (1968).
[32] *N. Grassie* and *H. W. Melville*, Disc. Faraday Soc., 2, 378, 1 (1947); Proc. Roy. Soc. Ser. A, 199, 14, 24 (1949).
[33] *F. Chevassus* and *R. de Broutelles*, "The Stabilization of Poly(vinylchloride)", E. Arnold, London (1963).
[34] *T. Kelen, G. Balint, G. Galambos* and *F. Tüdos*, Eur. Polym. J. 5, 597, 617, 629 (1969).
[35] *D. Braun* and *R. F. Bender*, Eur. Polym. J., Suppl. 229 (1969).
[36] *J. Voigt*, "Die Stabilisierung der Kunststoffe gegen Licht und Wärme", Springer, Berlin (1966).
[37] *K.-U. Bühler*, "Spezialplaste", Akademie-Verlag, Berlin (1978).
[38] *B. Doležel*, "Die Beständigkeit von Kunststoffen und Gummi", Hanser, München-Wien (1978).
[39] *K. Thinius*, "Stabilisierung und Alterung von Plastwerkstoffen", Verlag Chemie, Weinheim (1971).
[40] *W. W. Wright*, "The development of heat-resistant organic polymers" in *G. Geuskens* (ed.), "Degradation and Stabilization of Polymers", Appl. Science Publ., London (1975).
[41] *G. F. Pezdirtz* and *N. J. Johnston*, "Thermally stable macromolecules" in *R. S. Landell* and *A. Rembaum* (ed.), "Chemistry in Space Research", Elsevier, New York (1972).
[42] *V. V. Korshak*, Heat resistant polymers, Israel programme for scientific translations (1971).
[43] a) *E. Behr*, "Hochtemperaturbeständige Kunststoffe", Hanser, München (1969).
     b) *A. H. Frazer*, "High temperature resistant polymers", Interscience, New York (1968).
[44] *W. E. Gibbs* and *T. E. Helminiak*, in *A. D. Jenkins* (ed.), "Polymer Science", North-Holland, Amsterdam (1972).
[45] *M. M. Koton*, "High temperature polymers containing cyclic functions", in *W. M. Pasika* (ed.), Adv. Macromol. Chem. II, Academic Press, London and New York (1970).
[46] *W. Hale*, J. Polym. Sci. A-1, 5, 2399 (1967).
[47] *J. Preston*, J. Polym. Sci. A-1, 4, 2093 (1966).
[48] *S. S. Hirsch*, J. Polym. Sci. A-1, 7, 15 (1969).
[49] *E. L. Strauss*, "Polymer degradation processes in ablation" in *H. H. G. Jellinek* (ed.), "Aspects of Degradation and Stabilization of Polymers", Elsevier, Amsterdam (1978), p. 527.
[50] *G. F. D'Alelio* and *J. A. Parker* (eds.), "Ablative Plastics", Dekker, New York (1971).
[51] *D. L. Schmidt*, Report AFML-TR-66-78, May (1966).
[52] *J. R. Shelton*, "Stabilization against thermal oxidation" in *W. L. Hawkins* (ed.), "Polymer Stabilization", Wiley-Interscience, New York (1972).
[53] *W. L. Hawkins*, "The thermal oxidation of polyolefins — Mechanisms of degradation and stabilization" in *G. Geuskens* (ed.), "Degradation and Stabilization of Polymers", Appl. Science, London (1975).
[54] *J. C. Ambelang, R. H. Kline, O. M. Lorenz, C. R. Parks, C. Wadelin* and *J. R. Shelton*, Rubber Rev. Rubber Chem. Technol. 36, 1497 (1963).
[55] *H. C. Bailey*, "Chemistry of antioxidants, antiozonants and heat stabilizers" in *S. H. Pinner* (ed.), "Weathering and Degradation of Plastics", Columbine Press, Manchester (1966).
[56] *D. Braun*, "Recent progress in thermal and photochemical degradation of poly(vinylchloride)" in "Degradation and Stablization of Polymers", Appl. Science Publ., London(1975).

[57] *A. H. Frye* and *R. W. Horst*, J. Polym. Sci. 40, 419 (1959).

[58] *H. D. Beckey* and *H. R. Schulten*, "Field ionization and field desorption mass spectrometry in analytical chemistry" in *C. Merritt, C. N. McEwen* (eds.), "Practical Spectroscopy Series", Dekker, New York (1979).

[59] *G. J. Moll, R. J. Gritter* and *G. E. Adams*, "Mass spectrometry of thermally treated polymers" in *E. G. Brame Jr.* (ed.), "Application of Polymer Spectroscopy", Academic Press, New York (1978).

[60] a) *H. R. Schulten, H. D. Beckey, A. J. H. Boerboom*, and *H. L. C. Menzelaar*, Anal. Chem. 45, 2358 (1973).

     b) *H. R. Schulten*, "Pyrolysis field ionization and field desorption mass spectrometry of biomacromolecules, microorganisms and tissue material" in *C. E. R. Jones, C. A. Cramers* (eds.), "Analytical Pyrolysis", Elsevier, Amsterdam (1977).

[61] a) *I. Lüderwald* and *H. Ringsdorf*, Angew. Makromol. Chem. 29/30, 441 (1973);

     b) *I. Lüderwald* and *H. R. Kricheldorf*, Angew. Makromol. Chem. 56, 173 (1976).

[62] *J. Brandrup*, "Kunststoffe, Verwertung von Abfällen" in "Ullmanns Encyclopädie der technischen Chemie", 4th ed. Vol. 15, p. 411, Verlag Chemie, Weinheim (1978); Kunststoffe 65, 881 (1975); Ind. Anz. 99, 1535 (1977).

[63] *W. Kaminsky*, Ind. Anz. 99, 83 (1977).

[64] *W. Kaminski* and *H. Sinn*, Kunststoffe 68, 285 (1978); Ind. Anz. 99, 1605 (1977).

[65] *S. Speth*, Chem.-Ing. Techn. 45, 526 (1973).

[66] *D. F. Nicoli* and *G. B. Benedek*, Biopolymers 15, 2421 (1976).

[67] *A. Walton* and *J. Blackwell*, "Biopolymers", Academic Press, New York and London (1978).

[68] *J. Marmur* and *P. Doty*, J. Mol. Biol. 5, 109 (1962).

[69] *T. Tanaka* and *O. Vogl*, J. Macromol. Sci. Chem. A 8, 1071 and 1299 (1974); Polym. J. 6, 522 (1974); *M. Helbig, H. Inoue* and *O. Vogl*, J. Polym. Sci. 63, 329 (1978); Macromolecules 10, 1331 (1977).

3   Mechanical Degradation

3.1   Introduction

Mechanical degradation of polymers in its broader sense comprises a large field covering fracture phenomena as well as chemical changes induced by mechanical stress [1—15]. Because of their macromolecular chemical structure, plastics possess in many respects rather unique physical properties that account for their utilization as raw (engineering) materials for various purposes. Nevertheless, there exist load limits, i.e. the plastics engineer has to be aware of the loading capacity and of the fact that under permanent mechanical stress plastic materials might exhibit a behavior quite different from that encountered with metals and other inorganic raw materials.

In the past, sudden breakdown or fatigue phenomena resulting in morphological changes (deformations, crazing, cracks) have been treated predominantly on the basis of the physical nature of the development of defects. Only quite recently was attention drawn more generally to the chemical changes induced in polymers under the influence of mechanical stress. Such changes are documented in a great number of articles and in two recently published monographs [1, 2]. While "mechanochemistry" is certainly not a new chemical field, it has emerged but as a new method which appears to have interesting applications as discussed in more detail in a later section. We will attempt in this chapter to elucidate the role of chemical changes with respect to the development of physical defects and mechanical breakdown.

Before discussing various important aspects of mechanical degradation of polymers, it should be pointed out that under the influence of mechanical stress low molecular weight organic materials generally exhibit a different behavior to polymers. Normally, they do not undergo chemical changes if subjected to stress. Assemblies of low molecular weight molecules respond to applied stress by loosening intermolecular (physical) bonds resulting, macroscopically, in a deformation of the shape of the specimen and, microscopically, in a displacement of molecules. Thus, under the influence of shearing forces, intermolecular interactions between certain molecules at certain sites in a specimen are disrupted and new interactions become operative after the displacement. This holds not only for the liquid state but also for the solid state. Cracking crystals or amorphous specimens of low molecular weight compounds generally does not lead to the formation of free radicals, indicative for the scission of chemical bonds. On the other hand, free radicals are generally detected in polymers after mechanical treatment indicating ruptures of chemical bonds.

From this example it is evident that the mechanical degradation of polymers has to be looked upon not only from the viewpoint of the physicist or the plastic engineer but also from the sight of the chemist. In the following sections the present situation concerning the mechanism of mechanical degradation, including ultrasonic degradation, will be discussed. Subsequently, several kinds of application which demonstrate the rather general importance of mechanically induced changes in polymers will be described.

3.2   Mechanistic Aspects

3.2.1   Phenomenological Considerations

A variety of modes of imposition of stress onto polymers has to be considered with respect to mechanical polymer degradation.

As far as "pure" polymers (i.e. systems consisting almost exclusively of macromolecules) are concerned, machining (stirring, grinding, milling, mastication, processing in extruders etc.) and modifying processes (cutting, sawing, filing, drilling, free-cutting machining etc.) are important. Strain is also frequently imposed onto polymers when plastic articles are subjected to tensile or shear forces.

In all of these cases chemical bonds in the polymer chains can be ruptured. The extent to which bond scissions occurs, depends strongly on the state of the polymeric material. As far as materials consisting of linear macromolecules are concerned three states can be distinguished: a solid state (glassy or crystalline), a rubbery state (viscoelastic) and a molten state (elastoviscous). These states correspond to certain temperature regions with broad transition regions in between. In the case of thermosetting plastics, i.e. in crosslinked macromolecular systems, a molten state does not exist. On heating, these materials remain in the rubbery state up to their decomposition temperature.

The importance of the state can be inferred from the following simple concept. Mechanical energy transferred to a polymeric system can be dissipated via various relaxation processes in a harmless way, i.e. without inducing chemical changes. In competition with these relaxation processes, the scission of chemical bonds can occur. Obviously, the probability for bond scission should increase as the relaxation is impeded, i.e. more bonds should be ruptured upon increasing the rigidity of the material. However, as will be seen below, additional factors have to be considered with mixtures of polymers and low molecular weight compounds. Thus, even macromolecules in dilute solution degrade under the influence of stress, e.g. by high speed stirring and upon flowing through capillaries, as is described in Section 3.2.2. Ultrasonic degradation will be treated separately in Section 3.3.

A single, generally applicable mechanism of stress induced chemical reactions does not appear to exist. It seems that different bond scission mechanisms are operative depending on the state of the polymer and the mode of mechanical imposition of stress. In solid polymers, fracture planes and voids might give rise to the rupture of chemical bonds. In polymers in the rubbery or molten state or in solution, inter- and intra-chain entanglements might cause stretching of parts of the macromolecules, resulting eventually in bond scission.

At present it appears that strain is the prerequisite for bond rupture in polymer chains regardless of the state of the material, i.e. bond rupture occurs when sufficient energy is concentrated in a certain segment of a macromolecule as a consequence of the non-uniform distribution of internal stresses. Therefore, it seems reasonable that during the formation of a fracture plane, e.g. in a solid polymer where rapid slippage of polymer molecules is not possible, strains are generated which result in bond rupture.

3.2.2   Evidence for Bond Rupture

There are three kinds of evidence for scission of chemical bonds. The most direct approach is to carry out electron spin resonance (ESR) measurements. These permit the detection and, in many cases, also the identification of the free radicals generated by scission of bonds in the main chains or in side groups. Molecular weight determinations provide another direct, but less elegant way to demonstrate main-chain ruptures since the molecular weight of the polymer decreases as a result of main-chain degradation. Finally, a more indirect method for confirming bond rupture consists in the initiation of chemical reactions by reactive intermediates generated by scission of chemical bonds.

ESR spectra of various free macroradicals produced upon subjecting polymers to high stress [15—23] have been compiled by several authors [8—10]. Often, radicals generated directly by main-chain scission could be detected. Typical examples are shown in Table 3.1.

Table 3.1   Typical macroradicals generated by mechanically induced main-chain scission at 77 °K [15—23]

Polymer	Macroradicals
Polyethylene	$-\overset{\displaystyle H}{\underset{\displaystyle H}{C}}-\overset{\displaystyle H}{\underset{\displaystyle H}{C}}\cdot$
Polypropylene	$-\overset{\displaystyle CH_3}{\underset{\displaystyle H}{C}}-\overset{\displaystyle H}{\underset{\displaystyle H}{C}}\cdot \quad \cdot\overset{\displaystyle CH_3}{\underset{\displaystyle H}{C}}-\overset{\displaystyle H}{\underset{\displaystyle H}{C}}-$
Polyvinylalcohol	$-CH_2-\overset{\displaystyle H}{\underset{\displaystyle OH}{C}}\cdot \quad \cdot\overset{\displaystyle H}{\underset{\displaystyle H}{C}}-\overset{\displaystyle H}{\underset{\displaystyle OH}{C}}-$
Polytetrafluoroethylene	$-\overset{\displaystyle F}{\underset{\displaystyle F}{C}}-\overset{\displaystyle F}{\underset{\displaystyle F}{C}}\cdot$
Polymethylmethacrylate	$-\overset{\displaystyle H}{\underset{\displaystyle H}{C}}-\overset{\displaystyle CH_3}{\underset{\displaystyle R}{C}}\cdot \quad \cdot\overset{\displaystyle H}{\underset{\displaystyle H}{C}}-\overset{\displaystyle CH_3}{\underset{\displaystyle R}{C}}-$ $-\overset{\displaystyle CH_3}{\underset{\displaystyle R}{C}}-\overset{\displaystyle H}{\underset{\displaystyle R}{\overset{\cdot}{C}}}-\overset{\displaystyle CH_3}{\underset{\displaystyle \ }{C}}-\overset{\displaystyle H}{\underset{\displaystyle H}{C}}-H$ R: $\overset{\displaystyle \ }{\underset{\displaystyle O-CH_3}{C=O}}$

The radicals were frequently produced at 77 K by ball-milling or grinding in the absence of oxygen. A typical apparatus is shown in Fig. 3.1. Upon warming the samples or allowing O_2 (or other reactants) to enter the system, significant changes of the ESR spectra were observed, indicating radical rearrangement processes or reactions of the radicals with added substances.

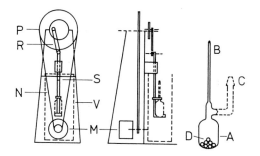

Fig. 3.1 Ball mill apparatus used for the generation of macroradicals in polymers at liquid nitrogen temperature (according to *Sohma* and *Sakaguchi* [8]).
A glass ampoule, *B* sample tube to be placed into the cavity of ESR spectrometer, *C* connector to vacuum system, *D* glass balls for milling, *M* motor, *N* belt, *P* pulley, *R* crank, *S* ampoule holder, *V* dewar flask containing liquid nitrogen

In many cases it turned out to be impossible to identify either one or both primary radicals generated by main-chain scission

$$
\begin{array}{ccc}
\text{H H} & & \text{H} \quad \text{H} \\
| \ | & & | \qquad | \\
-\text{C}-\text{C}- & \rightarrow & -\text{C}^{\bullet} + {}^{\bullet}\text{C}- \quad \text{(primary radicals)} \\
| \ | & & | \qquad | \\
\text{R H} & & \text{R} \quad \text{H}
\end{array}
\tag{3.1}
$$

because these radicals are extremely unstable even at 77 K. For example, in the case of polybutadiene, the ESR spectrum of the primary radical

$$
\begin{array}{cc}
\text{H} & \text{H} \\
| & | \\
{}^{\bullet}\text{C}-\text{C}=\text{C}-\text{C}- \\
| \quad | \ | \ | \\
\text{H} \ \text{H} \ \text{H} \ \text{H}
\end{array}
$$

PB$-$I$^{\bullet}$

could not be detected, because it is converted easily to a secondary radical by a shift of a hydrogen atom [8].

$$
\begin{array}{cc}
\text{H} & \text{H H} \\
| & | \ | \\
{}^{\bullet}\text{C}-\text{C}=\text{C}-\text{C}-\text{C}- \\
| \ | \ | \ | \ | \\
\text{H H H H H}
\end{array}
\rightarrow
\begin{array}{cc}
\text{H} & \text{H} \\
| & | \\
\text{H}-\text{C}-\overset{\bullet}{\text{C}}-\text{C}=\text{C}-\text{C}- \\
| \ | \ | \ | \ | \\
\text{H H H H H}
\end{array}
\tag{3.2}
$$

PB$-$I$^{\bullet}$ 　　　　　　　PB$-$II$^{\bullet}$

The radical PB—II· could be identified [8]. Similarly, the radical

$$
\begin{array}{ccccc}
& CH_3 & H & CH_3 & H \\
& | & | & | & | \\
-C & \!\!-\!\!- & C-C & \!\!-\!\!- & C-H \\
& | & \bullet & | & | \\
& R & & R & H
\end{array}
$$

could be identified as a secondary radical in the case of PMMA (see Table 3.1).

Generally, reactions of primary or secondary radicals with molecular oxygen or other reactants can be detected rather easily. Fig. 3.2a shows the ESR spectrum obtained with polypropylene after ball milling at 77 K in vacuo. It has been interpreted [8] as resulting from the overlapping of an octet due to the radical $-CH_2-\overset{\bullet}{C}-CH_3$ and a

quartet due to the radical

$$
\begin{array}{ccc}
CH_3 & H \\
| & | \\
-C & \!\!-\!\!- & C\!\cdot \\
| & | \\
H & H
\end{array}
$$

The spectrum obtained after bringing the sample

into contact with O₂ (see Fig. 3.2b) indicates the coexistence of the radical

$$
\begin{array}{ccc}
H & & \\
| & & \\
-C-\overset{\bullet}{C}-CH_3 \\
| & | \\
H & H
\end{array}
$$

and the peroxyl radical

$$
\begin{array}{ccc}
CH_3 & H \\
| & | \\
-C & \!\!-\!\!- & C-O-O\!\cdot \\
| & | \\
H & H
\end{array}
$$

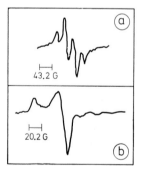

Fig. 3.2   Evidence for free radicals generated by ball milling of polypropylene at 77°K.
(a) ESR spectrum observed after milling in vacuo, (b) ESR spectrum observed after bringing the sample into contact with oxygen

Molecular weight changes induced by stress were studied systematically mostly with polymer solutions (see Section 3.2.4) and only rarely with neat polymers. In earlier investigations properties related to the molecular weight (e.g. the intrinsic viscosity) were recorded as a function of time of mechanical treatment, only recently have molecular weight or molecular weight distribution changes been measured directly. Typical results are presented in Fig. 3.3 and Fig. 3.4. The decrease of the average molecular weight of polystyrene as a function of ultrasonic irradiation time is shown in Fig. 3.3 [40]. The irradiation was carried out at two different polymer concentrations, the extent of degradation being more pronounced in the more dilute solution.

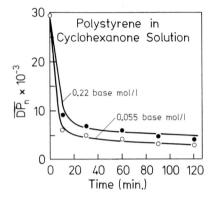

Fig. 3.3 Ultrasound-induced main-chain scission of polystyrene in N_2 saturated cyclohexanone solutions at 60 °C and at two different concentrations (as indicated on the graph). Number average molecular weight (determined osmometrically) vs. the time of ultrasonic treatment [40]

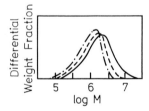

Fig. 3.4 Shear degradation of polyisobutene. Changes in MWD during repetitive extrusion of polyisobutene in a capillary rheometer at 80 °C. Number of capillary passes: (————) 0; (– – – –) 5; (– · –) 10 [24]

Fig. 3.4 shows molecular weight distribution curves of polyisobutene obtained after extrusion of the polymer in a capillary rheometer at 80 °C [24]. Shear degradation causes the average molecular weight to decrease and the molecular weight distribution to become narrower ($\overline{M}_{w,0} = 1.8 \, \overline{M}_{n,0}$; after 10 passes $\overline{M}_w = 1.5 \, \overline{M}_n$). This example demonstrates that mechanical main-chain scission is, in this case, a non-random process. It has been shown in many studies of mechanical degradation of linear macromolecules, that the probability for main-chain rupture is highest around the middle of the chain (see Section 3.2.3).

A phenomenon pertaining to stress-induced main-chain scission is the so-called limiting degree of polymerization. As shown in Fig. 3.3, the average molecular weight changes exponentially with time of application of stress and approaches a limiting value at prolonged times of treatment. It has been assumed, therefore, that the chain stability increases with decreasing chain length. This implies that plots of $\overline{\mathrm{DP}}_n$ or $(\overline{\mathrm{DP}}_n)^{-1}$ vs. time of treatment do not lead immediately to mechanochemical yields of bond scission comparable, for instance, to quantum yields in the case of photochemical main-chain scission (see Chapter 4). With respect to main-chain scissions, as evidenced by molecular weight measurements, it can be concluded that, at present, there is still a lack of suitable experiments allowing direct quantitative interpretations of the results. However, on a qualitative basis, it can be quite clearly inferred from a great number of molecular weight measurements that in polymers, under the influence of stress, chemical bonds are ruptured.

The great number of patents and articles [1, 11 – 14, 25, 26], devoted to the still growing field of initiation of chemical reactions by stress-generated reactive intermediates, can be taken as evidence for chemical bond rupture. Here, however, attention is drawn only to one example which is typical of mechanochemical (or tribochemical) processes. It concerns the formation of block-copolymers via macroradicals ($-AAAAA\cdot$) which react with a monomer B when the polymer ($-AAAAAA-$) is subjected to stress in the presence of the monomer:

$$-A-A-A-A-A- \xrightarrow{\text{stress}} 2-A-A-A-A\cdot \qquad (3.3)$$

$$-A-A-A-A\cdot + nB \rightarrow -A-A-A-A-B-(B)_{n-2}-B\cdot \qquad (3.4)$$

Additional examples will be presented in Section 3.4.5.

It is generally accepted that homolytic bond scission occuring in polymers under stress leads to free radicals. Heterolytic bond scission, leading to macroions, has been demonstrated with inorganic solids e.g., silicates and asbestos [27–31]. The stress-induced formation of ionic intermediates in synthetic organic polymers has also been discussed occasionally [32–35].

3.2.3   Theoretical Aspects of Bond Rupture and Experimental Findings

The previous sections were devoted to the fact that chemical bonds are ruptured under the influence of stress and to the question of how they are ruptured. In this section two further questions will be considered: (a) Why are chemical bonds broken when a polymer is subjected to stress? (b) Are fracture and deformation correlated with bond rupture? With respect to question (a) it must be pointed out that the critical conditions for bond rupture depend both on the amount of "elastic" energy stored in a single macromolecule and on the time the macromolecule remains in the mechanically excited, i.e. the strained, state. Relevant amounts of energy must, of course, be equal to or greater than the bond dissociation energy. As far as can be concluded from the few quantitative determinations published so far, only a very small fraction of the "elastic" energy absorbed by the polymer results in bond scission. The energy required to rupture at 80 °C one mol of bonds in polyisobutene, for instance, was determined as ca. 8.4×10^6 kJ [24] (bond dissociation energy 330 kJ/mol). Therefore, bond scission is a rather rare event relative to other energy dissipation processes. The non-chemical relaxation processes comprise slippage of chains relative to surrounding molecules (enthalpy relaxation) and changes of chain conformation (entropy relaxation). The various aspects of mechanical excitation and de-excitation of macromolecules have been discussed by Kausch [36]. In kinetic terms, harmless, i.e. non-chemical, relaxation processes compete with bond rupture. As these relaxation processes frequently occur rather rapidly, the non-chemical relaxation time is a controlling factor with respect to bond rupture. If the time for bond scission is considered constant, an increase in (non-chemical) relaxation time should lead to a higher yield for bond rupture and vice versa.

In the past, various theoretical approaches have been developed with the aim of elucidating the occurrence of bond ruptures in amorphous polymers at temperatures above the glass transition temperature and in polymers in dilute solutions. An important approach was developed by *Frenkel* [37], and *Kauzmann* and *Eyring* [38] who assumed that under the influence of shear individual linear macromolecules are extended in the direction of stress. Thereby the bonds in the middle of the chains are strained while the remainder of the macromolecules is almost uneffected. No degradation was expected for macromolecules with $DP < DP_{lim}$ and at shear rates below a limiting value. A theory, presented by *Bueche* [39], concerns systems containing polymers in more or less close contact with each other (rubbery systems, melts and concentrated solutions). It predicts that entanglements induce preferentially tensions in the middle portion of the chains and thus make bonds located in the center of the chains more succeptible to breakage. Furthermore, the rate of main-chain rupture is assumed to increase significantly with increasing molecular weight.

Thus, mechanical degradation would appear to be a non-random process, a fact which should be detectable by typical molecular weight distribution (MWD) changes as outlined in Chapter 1.

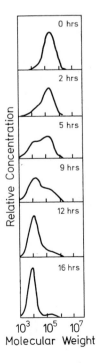

Fig. 3.5   Main-chain scission of polystyrene upon milling. Gel permeation chromatograms indicating the change of MDW during milling (time of treatment in hours as indicated on the graphs) [41]

Only rather recently, the advent of the GPC method has made it possible to investigate systematically MWD changes as a function of the degree of degradation. It was found in several cases that either a bimodal distribution resulted or that an initial random MWD became narrower upon main-chain scission. In other cases, the results indicated a random main-chain rupture mechanism. Typical results for both cases are presented in Fig. 3.4, Fig. 3.5, and Fig. 3.6. From the observed MWD changes it may be concluded that mechanical stress can induce non-random main-chain rupture in linear polymers in agreement with the above mentioned theories. No definite answer to the question of why in certain cases random chain scission occurs can be given at present. However, it appears [1], that the mechanism of main-chain scission depends on the rates of mechanical excitation and de-excitation of chains, the probability and the extent of entanglement formation and the interaction of the macromolecules with the surrounding matrix.

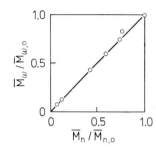

Fig. 3.6   Ultrasonic degradation of polyiso-butene in 1 . 2, 4-trichlorobenzene solution (90 g/l) $\overline{M}_{w,0} = 2\overline{M}_{n,0}$ [42]

The latter interaction pertains especially to slippage phenomena. The distribution of scissions along the chains depends to a large degree on whether or not mechanical stress is imposed on individual macromolecules or on a more or less compact array of macromolecules. This aspect will be treated in more detail in Section 3.2.4 in connection with degradation of polymers in solution. As far as present knowledge is concerned, mechanical treatment of linear polymers can lead to non-random main-chain scission under conditions favoring the straining of central bonds. On the other hand, if strains are generated in various parts of a chain simultaneously, random main-chain scission is favored.

It appears to be appropriate to discuss here the phenomenon of the limiting degree of polymerization in more detail. The existence of $\overline{DP}_{lim}$ has been demonstrated for various polymers and for various modes of mechanical treatment (mastication of rubbery polymers, extrusion of polymer melts, high-speed stirring and ultrasonic irradiation of polymers in solution). Confusion arose about $\overline{DP}_{lim}$, since reported values frequently, were at variance.

In order to understand that $\overline{DP}_{lim}$ is not a natural constant characteristic for each polymer, it has to be recalled that a prerequisite for bond scission is the capability of macromolecules to absorb and store mechanical energy for a time sufficiently long to achieve bond scission. The capability of absorbing and storing mechanical energy depends on a number of variables, one of them being the molecular weight. The following may serve as an aid to comprehension. In a system of low viscosity, the interaction of a macromolecule with surrounding molecules is relatively weak, i.e. the probability of absorbing energy is low. As de-excitation, on the other hand, is rather effective, the capability of storing energy is also low. Therefore, a $\overline{DP}_{lim}$-value results which is higher than that pertaining to a highly viscous system.

It should be pointed out that only careful and further systematic studies can lead to the elucidation of problems concerning $\overline{DP}_{lim}$.

Considerations, so far, have concentrated on polymeric systems of relatively high inner mobility and a high degree of homogeneity. In the following discussion more rigid systems will be dealt with. For practical applications, partially crystalline polymers (e.g. polyethylene) are of eminent importance. With respect to the mechanical degradation of these polymers the reader's attention will now be directed to a model developed by *Peterlin* [20−22] and to relevant experimental results reported by *Sohma* et al. [8]. If stress is applied to a polymer specimen consisting of amorphous and crystalline regions and if the mechanical force imposed is gradually increased, initially main-chain bonds will be ruptured almost exclusively in the amorphous regions connecting the crystalline regions by "tie molecules". As schematically depicted in Fig. 3.7, the shortest tie molecules will be ruptured primarily. Additional tie molecules will be broken upon increasing the strain.

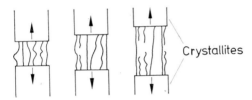

Fig. 3.7   Schematic illustration of bond rupture in the amorphous regions of semi-crystalline polymers. (After *Peterlin* [21])

Experimentally, this assumption was verified by the finding that in polyethylene the concentration of free radicals increased with increasing strain. If the amorphous parts were removed by treating the polymer with HNO_3 prior to ball milling, free radicals could not be detected, which implies also that rather high mechanical forces are needed in order to accomplish bond scissions in crystalline regions of polymers.

In order to elucidate the question of how intermolecular forces influence the bond scission efficiency, linear hydrocarbons of $\overline{DP}_n$ varying between 16 and 136 were exposed to ball mill treatment [8]. Free radicals were detected only at chain lengths above about 70 to 100. A critical degree of polymerization of this magnitude becomes plausible upon assuming that bond rupture in the main-chain is only possible if

$$E_{C-C} \le \overline{DP}_{crit} E_i \tag{3.5}$$

where E_{C-C} and E_i designate the bond dissociation energy and the interaction energy of the repeating unit with the surrounding molecules, respectively. Thus, for polyethylene, $\overline{DP}_{crit} \approx 80$ is obtained (with $E_i = 4.2$ kJ/mol, estimated from the activation energy of flow and $E_{C-C} = 343$ kJ/mol). This corresponds fairly well to the experimentally determined critical chain length range.

Concerning the question as to whether or not fracture and deformation are correlated with bond rupture, a process of clarification is in progress at present and seems to be approaching its final stages. The subject has recently been discussed at length by *Kausch* [43]. To an increasing extent it appears that arguments are being gathered that speak against the hypotheses correlating fracture and deformation with chain breakage. The discussion of this interesting problem was initiated by a hypothesis of *Zhurkov* et al. [44—46], who, whilst investigating the formation of so-called submicrocracks, detectable by *X*-ray scattering, in polyethylene and polypropylene found, that the concentration of end groups, indicating main-chain scission, was several orders of magnitude greater than the concentration of free radicals. It was assumed that submicrocracks can be generated by individual radical pairs which induce a chain reaction involving the surrounding macromolecules:

$$\sim\!\sim\!\sim\!\sim\!\sim \longrightarrow \sim\!\sim\!\sim\!\cdot \; + \; \cdot\!\sim\!\sim\!\sim \tag{3.6}$$

$$\sim\!\sim\!\cdot \atop \sim\!\sim\!\sim\!\sim\!\sim \longrightarrow \sim\!\sim \atop \sim\!\sim\!\cdot\!\sim\!\sim \tag{3.7}$$

$$\sim\!\sim\!\dot{}\!\sim\!\sim \longrightarrow \sim\!\sim\!\cdot \; + \; \sim\!\sim \tag{3.8}$$

In the case of polyethylene reaction (3.8) would correspond to the reaction:

$$R-\underset{\cdot}{C}H-CH_2-R' \rightarrow R-CH=CH_2 + \cdot R' \tag{3.9}$$

The spontaneous decay of lateral macroradicals according to reaction (3.9) was assumed by *Zakrevskii* and *Zhurkov* [44—46] to be favored as a consequence of a diminution of the activation energy under the influence of stress. Whereas this assumption appears to be dubious, the origin of submicrocracks was explained by *Peterlin* [47] by a stress-induced retraction of the ends of microfibrils. If the latter are thought to be located preferentially at the outer surface of the fibrils, a retraction leads to the formation of voids. A recent re-investigation [48] of this problem, involving a comparison of free radical and main-chain scission yields, led to the conclusion that the hypothesis correlating microcrack formation strongly with main-chain scission is highly questionable.

3.2.4   Polymer Degradation in Solution

Stress-induced reactions can occur if polymer solutions are subjected to high-speed stirring, shaking, turbulent flow or ultrasonic treatment. Problems caused by turbulent flow of dilute polymer solutions will be dealt with below in connection with drag reduction (see Section 3.4.2). Ultrasonic degradation will be discussed in the following section.

A great number of synthetic and natural linear polymers have been investigated in solution in various solvents, e.g. polystyrene, poly-α-methylstyrene, polymethylmethacrylate, polyisobutene, polyvinylacetate, poly(ethylene oxide), polyacrylamide, polyvinylpyrrolidone, polyvinylalcohol, DNA, poly(L-glutamic acid) and poly-L-tyrosine. The subject has been reviewed recently by *Casale* and *Porter* [1] and relevant references are to be found there.

For systematic studies, stress-induced reactions were frequently induced by forcing polymer solutions to flow through capillaries or by treating polymer solutions in rotational viscometers. For high-speed stirring studies homogenizers have been used.

Although most studies were aiming at detecting main-chain scissions, unfortunately only few researchers attempted the direct determination of these scissions. Intrinsic viscosity measurements usually served to arrive at conclusions concerning the influence of polymer concentration, solvent power, shear rate, shear stress and temperature on polymer degradation. At present it appears that the general dependence of shear degradation on these parameters is not well understood, as results obtained in different laboratories using different instruments are in many cases at variance.

At low concentrations and constant shear stress, the extent of degradation was found higher at lower concentration for polyisobutene [49], polymethylmethacrylate and polycaprolactam [50, 51]. This might be due to the fact that intermolecular entanglement becomes less important as the concentration decreases. Intermolecular entanglement may in dilute solution provide more rapid modes of de-excitation, i.e. the dissipation of input energy might be facilitated and, thus, the extent of degradation might be reduced. This view was corroborated by the finding that, in the case of poly-α-methylstyrene in solution, degradation is independent of polymer concentration below a critical concentration (0.15 g/l in this case) [52]. In concentrated solutions intermolecular entanglement can play a quite different role insofar as macromolecules are prevented from responding as a unit to shear forces and, therefore, C—C bonds in the main-chain are ruptured. *Fukutomi* et al. [53] contributed towards the problem of concentration dependence of shear degradation by a careful GPC analysis of poly-α-methylstyrene degraded in toluene solution upon flowing through a capillary. Fig. 3.8 shows typical results obtained at a concentration of 1 g/l. The MWD becomes narrower as the average molecular weight decreases due to continued main-chain degradation.

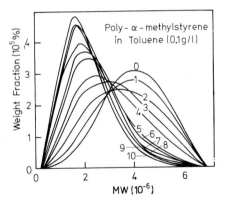

Fig. 3.8 Mechanical degradation of poly-α-methylstyrene in toluene solution, at room temperature induced by flowing through a capillary. Differential MWD as measured before treatment (O) and after various passes through the capillary (the number of passes is indicated at the curves). Polymer concentration: 1 g/l. $\overline{M}_{w,0}$: 4.4×10^6. *Fukutomi* et al. [53])

Following the theory of *Simha* [54], the scission probability of a base unit (expressed as a first order rate constant k) was calculated as a function of the location in the chain. Fig. 3.9 shows typical results. The numbers at the curves are a measure for the chain length (decreasing with decreasing chain length). It can be seen that, at high chain lengths, k is highest at the center of the chains and decreases with increasing distance from the center. Thus, the non-random character of main-chain scission is evident. Fig. 3.9 shows, furthermore, that the randomness of main-chain degradation increases with decreasing chain length. In other words, for shorter chains the probability for breaks in the middle portion of the chain is much smaller than for longer chains. The analysis of results obtained at higher concentrations (up to 3 g/l) revealed the following: as the polymer concentration increases, the rate constant k becomes generally smaller and much less dependent on the location of the respective base unit in the chain. This implies that random chain cleavage is approached with increasing polymer concentration.

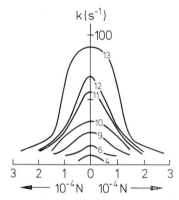

Fig. 3.9 Mechanical degradation of poly-α-methylstyrene in toluene solution at room temperature. Probability k of chain scission as a function of the distance N between site of rupture and center of the chain. The numbers at the curves correspond to the MW (decreasing MW with decreasing number). (After *Fukutomi* et al. [53])

With respect to solvent power, there are indications that stress-induced main-chain scission occurs more efficiently in poor solvents than in good solvents [52, 55, 57, 58]. For polymers in solution, the dependence of shear degradation on shear rate has scarcely been investigated and, therefore, no attempt will be made here to arrive at general conclusions. In various cases it was found that the extent of degradation increased with increasing applied shear stress [49—51, 58—61].

3.3   Ultrasonic Degradation

3.3.1   The Importance of Ultrasonic Degradation

If polymers in solution are subjected to the influence of ultrasonic waves, quite generally, main-chain degradation occurs. With the aid of ultrasonic waves mechanical energy can be conveniently dispersed in polymer solutions. It appears furthermore, that, with respect to reproducibility and speed, main-chain degradation of polymers can be achieved by this method more efficiently than by high-speed stirring or capillary flow treatment. Owing to this outstanding capability of inducing main-chain rupture and because *cavitation* is a prerequisite for mechanical excitation of macromolecules in solution, ultrasonic degradation is treated separately here. The technique of generating ultrasound has been developed to a high standard during the last decades. At present ultrasonic generators are commercially available or can be assembled relatively easily. With the aid of these generators, high molecular weight polymer samples can be converted to low molecular weight material. Another application concerns the production of polymer fractions of relatively narrow MWD. Most applications of the ultrasonic degradation technique, however, pertain to the fields of biochemistry and molecular biology where ultrasound is quite commonly employed to disrupt cells and tissues.

The field of ultrasonic degradation of polymers in solution has been reviewed recently [6, 62].

3.3.2 Experimental Methods

Ultrasonic waves are commonly generated with the aid of vibrators which act either as piezoelectric or as magnetostrictive oscillators. For the generation of high intensity ultrasound, quartz, barium titanate and lead zirconate/titanate are used as materials for piezoelectric oscillators. Nickel and iron/cobalt alloys are used for magnetostrictive oscillators. Depending on the frequency and the material, up to 90% of the applied electrical energy can be converted to ultrasonic energy. Intensities of up to 500 W/cm² are attainable. Frequencies range from 20 kHz to about 1 MHz, corresponding to a wavelength in water of 7.5 to 0.15 cm. Owing to the strong attenuation of transverse waves in liquids, only longitudinal waves are operative with respect to the induction of chemical changes.

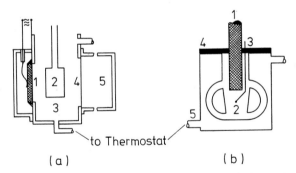

Fig. 3.10   Experimental devices for the study of ultrasonic degradation of polymers in solution. (After *Basedow* and *Ebert* [6]).
(a) Set-up with transducer separated from sample cell.
(1) vibrator, (2) sample cell (glass), (3) paraffin oil, (4) membrane, (5) absorber.
(b) Set-up with transducer dipping into the polymer solution.
(1) vibrator, (2) polymer solution, (3) thermo couple, (4) cover lid, (5) connection to thermostat

Typical set-ups used for polymer degradation studies are shown in Fig. 3.10 (a) and (b). The transducer (vibrator) is in both cases in direct contract with an electrically insulating liquid. In Fig. 10 (b) the transducer dips into the polymer solution which is forced to flow through the side arms. Details pertaining to the generation of ultrasonic waves have been reported in the literature [63, 64].

3.3.3   Mechanism of Ultrasonic Degradation

Cavitation and direct effects.

Upon subjecting a liquid, saturated with air or another gas, to ultrasonic irradiation, small bubbles are produced, a phenomenon related to *cavitation*. When the liquid is exposed to ultrasonic waves of high intensity it expands and gives rise to a negative pressure causing the dissolved gas to form bubbles. Assuming an intensity of 10 W/cm² and a frequency of 0.5 MHz, the displacement amplitude is estimated as 1.2×10^{-2} cm corresponding to a pressure amplitude of 5.4 bar. On this basis it was concluded that the onset of cavitation corresponds to a pressure amplitude of ca. 1 bar [6].

It should be emphasized that "genuine" cavitation can occur only at pressure amplitudes above 100 bar, a value far beyond realization with the commonly used ultrasonic generators. Therefore, the use of the term cavitation for the ultrasound-induced formation and collapse of bubbles in liquids containing dissolved gases is somewhat incorrect.

A *direct action* of ultrasonic waves on macromolecules in solution via "genuine" cavitation can be excluded. Furthermore, older conceptions, assuming other modes of direct action [65], do not yield a general explanation of the experimental findings. Therefore, at present, direct action is considered to play only a minor role (if any at all) during ultrasound-induced main-chain rupture.

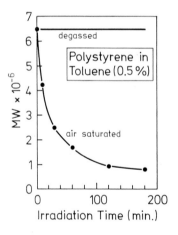

Fig. 3.11 Ultrasonic degradation of polystyrene in toluene solution at ambient temperature. Average molecular weight vs. time of irradiation. Polymer concentration: 1%; frequency: 0.4 MHz; intensity: 50 W. (After *Weissler* [66])

The importance of the presence of gas for the generation of bond ruptures has been clearly demonstrated. Typical results [66] are presented in Fig. 3.11, which shows that polystyrene undergoes main-chain scission only in the presence of air and not in a degassed solution. The formation of bubbles is a rather complicated process which will not be discussed here. It should only be noted that there is a nucleation phase where small bubbles are formed which combine to give bigger bubbles. The latter can exist for a relatively long period, during which time their size and geometry change (oscillating cavities), before eventually collapsing. The adiabatic collapse of a gas bubble gives rise to a pressure wave (shock wave) [67–69]. It has been shown [80] that shock waves, of approximately spherical chape, are radiated from the bubble at the moment of implosion. Main-chain rupture in macromolecules is thought to be induced by these shock waves [71, 72], which are assumed to cause a rapid compression with subsequent expansion of the liquid. On the molecular level, this implies a rapid motion of solvent molecules to which the macromolecules embedded in the solvent cannot adjust. Thus, friction is generated which causes strain and eventually bond rupture in the macromolecules.

Apart from the action of shock waves, the collapse of cavitation bubbles is thought to create pronounced perturbations in the surrounding liquid. The ensuing high velocity gradient generates flow fields, i.e. induces inhomogeneous flow of the solvent molecules. If a macromolecule, or part of it, is located in a region of a high velocity gradient and if it is sufficiently strained, chain fractures become feasible. Relevant theories have been developed since 1954 [73 — 76].

3.3.4   Degradation Studies

The action of ultrasonics on polymers in solution was studied in two ways: one aiming at the detection of transient species via chemical reactions, the other concerning main-chain scissions and ensuing MW and/or MWD changes. For the qualitative and quantitative detection of free macroradicals their reactions with radical scavengers or stable

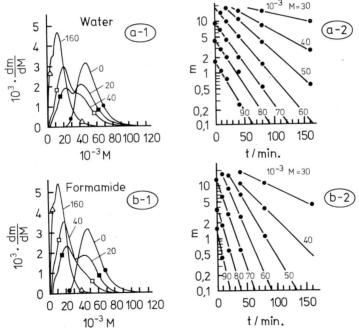

Fig. 3.12 Ultrasonic main-chain degradation of dextran in water and formamide solution at 20°C and 25 °C, resp. (a—1) and (b—1) differential molecular weight distributions. Numbers at the curves denote time of irradiation. (a—2 and (b—2): weight fraction m vs. time of irradiation. Numbers on the curves denote the MW. Polymer concentration: 5 g/l.

$\overline{M}_{w,0} = 5.32 \times 10^4$; $\overline{M}_{n,0} = 4.92 \times 10^4$. Ultrasonic intensity: 24 W cm^{-2}, frequency: 20 kHz. (After *Basedow* and *Ebert* [80]) free radicals, such as DPPH, were studied [77]. These studies also yielded information concerning the kinetics of main-chain degradation. Further evidence for the generation of free macroradicals was derived from the fact that the radical polymerization of unsaturated monomers is initiated upon subjecting polymers to ultrasonic irradiation in the presence of monomers [78, 79].

Typical results pertaining to MW measurements have already been given in Fig. 3.3 and 3.6. Some additional typical data, obtained recently will be presented here. Fig. 3.12 shows differential molecular weight distribution curves of dextran that had been subjected to ultrasonic irradiation for various times in dilute solution (solvents: water and formamide) [80]. It can be seen that main-chain degradation causes a diminution of the maximum, pertaining to the unirradiated polymer, with the simultaneous formation of a new maximum, corresponding to a MW about 50% smaller than the initial MW. This indicates that the center parts of the macromolecules are preferentially prone to bond ruptures. Furthermore, from this work [80] it was concluded, that the main-chain cleavage process follows a 1st order law, the rate constant increasing linearly with increasing MW.

Other investigations [81] were devoted to the question of whether or not a limiting degree of polymerization exists below which main-chain scission does not occur. Fig. 3.13

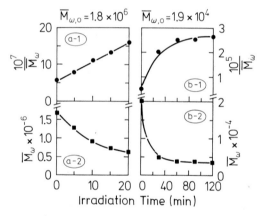

Fig. 3.13 Ultrasonic main-chain degradation of polystyrene in tetrahydrofuran solution (20 g/l) at $(10 \pm 3)\,°C$. Frequency 20 kHz; initial molecular weight: 1.8×10^6 (a) and 1.9×10^4 (b).
Plots of MW and $(MW)^{-1}$ vs. time of irradiation. (After *Sheth, Johnson* and *Porter* [81])

shows the results. Polystyrene samples of rather different initial MW (1.8×10^6 and 2×10^4) were subjected in solution to ultrasonic irradiation. In the case of the high molecular weight sample a plot of $1/\overline{M}$ vs. irradiation time yields a straight line indicating that the number of main-chain scissions per base unit is directly proportional to the time of irradiation. This result, furthermore, leads to the conclusion that the rate of energy uptake does not depend on the MW as long as the latter is rather high. In the case of the low molecular weight sample this does not hold (as can be seen from Fig. 3.13 (b-1)). It is, however, remarkable that main-chain rupture occurs at all in this case, as the initial MW is lower than limiting MW values reported earlier for polystyrene [82]. It should, be pointed out here again, therefore, that stress-induced main-chain rupture depends essentially on the rate of energy uptake by the polymer, and the competition between the various processes of de-excitation, including the chemical route via bond scission.

Insofar as it depends on various parameters, e.g. chain length, extent of interaction with the solvent and, in the case of flexible linear macromolecules, on coil density, the rate of energy uptake appears to be a determining factor with respect to the limiting degree of polymerization.

3.4   Applications

Applications of mechanical degradation cover a wide range from deleterious effects to the synthesis of new polymers. In the following paragraphs typical examples are presented. The general importance of stress-induced main-chain scission combined with subsequent oxidation reactions in the processing of thermoplastic polymers is treated in Section 3.4.1. Subsequent sections (3.4.2 and 3.4.3) are devoted to more special cases concerning deleterious actions of stress. Finally, beneficial aspects are discussed in Sections 3.4.4 and 3.4.5 dealing with mastication and mechanochemistry.

3.4.1   Stress-induced Chemical Alterations of Polymers

Normally, commercial polymers are not chemically pure i.e. their chemical composition does not correspond exactly to the formula of the base unit. Apart from admixed impurities, polymers frequently contain very small amounts of substituents chemically linked to the macromolecules. A possible source for such "chemically incorporated impurities" may be stress-induced chemical alterations.

Generally chemically incorporated impurities are of minor importance in network polymers (duromers and elastomers), however, they play an important role in most linear polymers.

The chemical nature of the impurities can vary widely. As far as environmental stability is concerned, unsaturated $C-C$ bonds and oxygen containing groups, e.g., carbonyl, peroxide or hydroperoxide groups, can be of importance since they are capable of absorbing light and of starting chemical reactions.

If linear polymers are subjected to mechanical treatment during thermoplastic processing, main-chain scissions can be induced, e.g.:

$$\sim CH_2-CH_2-CH_2\sim \rightarrow \sim CH_2^{\cdot} + \cdot CH_2-CH_2\sim \qquad (3.10)$$

The free radicals produced in reaction (3.10) can disproportionate yielding unsaturation:

$$\sim CH_2^{\cdot} + \cdot CH_2-CH_2\sim \rightarrow \sim CH_3 + CH_2=CH\sim \qquad (3.11)$$

Frequently, as oxygen is present during processing, the macroradicals can react according to reaction (3.12):

$$\sim CH_2^{\cdot} + O_2 \rightarrow \sim CH_2-O-O\cdot \qquad (3.12)$$

which might be followed by the formation of hydroperoxide groups via reaction (3.13)

$$\sim\!CH_2\!-\!O\!-\!O\cdot \quad + \quad \overset{\textstyle\langle}{\underset{\textstyle\rangle}{CH_2}} \quad \longrightarrow \quad \sim\!CH_2\!-\!O\!-\!OH \quad + \quad \overset{\textstyle\langle}{\underset{\textstyle\rangle}{\cdot CH}} \qquad (3.13)$$

Reaction mechanisms elucidating the formation of other oxygen containing groups are dealt with elsewhere in this book (see Sections 1.3.2 and 7.4.4). Obviously, mechanically induced main-chain rupture is a means of initiating oxidative chain reactions, i.e. autoxidations of polymers, as discussed before in Section 2.3 in connection with thermal degradation of polymers. Stress-induced main-chain rupture might also initiate the depolymerization of linear polymers (see chapter 2 and reaction (2.4) in Scheme 2.1).

In conclusion, two aspects must be pointed out with respect to the deleterious action of mechanical forces on linear polymers during thermoplastic processing: (a) Stress-induced free radical chain reactions can be initiated causing chemical alterations with the consequence of changes in the physical properties of the polymer. (b) Chemical impurities can be incorporated into the macromolecules, which might give rise to the initiation of chemical reactions at a later period, e.g. on exposing the polymer to sunlight.

3.4.2   Drag Reducers and Viscosity Index Improvers

An interesting application of synthetic polymers concerns the reduction of drag in turbulent flow of liquids. On pumping liquids through tubes, the rate of flow is strongly influenced by vortices generated by interactions of the liquids with the walls of the tubes. High flow rates are necessary to transport liquids economically over long distances through pipe lines. It was found that the flow rate can be significantly increased upon addition of minute amounts of flexible linear polymers to the liquid [83]. Other materials capable of acting as drag reducers are: colloidal soap solutions, suspensions of fine insoluble particles or fibers (clay, asbestos etc.) with long-chain samples of poly(ethylene oxide) and polyacrylamide; a remarkable drag reducing efficiency was reported for aqueous solutions containing less than 10 ppm of added polymer [84—86]. The phenomenon of drag reduction is based on a shift to higher values of the critical Reynolds number *) for the transition from laminar to turbulent flow.

Applications of drag reducers [87] pertain to an increased capacity of pipe line and hose systems for pumping liquids (e.g. water in fire-fighting hose lines and crude oil pipe lines).

The importance of polymers as drag reducers has limitations due to the fact that drag reducing is combined with main-chain degradation. The drag reducing efficiency, on the other hand, depends strongly on the molecular weight of the polymer, as demonstrated by the data for poly(acrylic acid) presented in Fig. 3.14 [88], which may serve as a typical example. In Fig. 3.14 the relative drag, as a measure of the drag reducing efficiency, is plotted as a function of the polymer concentration (relative drag = square of the ratio of the flow time of the solution to that of water). At very low polymer concentrations the relative drag decreases with increasing polymer concentration. After

---

*) Reynolds number: $R_e = \dfrac{\overline{V}\,d\varrho}{\eta}$

$\overline{V}$: average velocity, d: tube diameter, ϱ: liquid density, η: viscosity

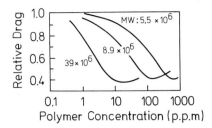

Fig. 3.14 Relative drag at constant shear stress vs. concentration of poly(acrylic acid) of MW as indicated at the curves. Neutral aqueous solutions. Relative drag $= (t_{solution}/t_{water})^2$, ($t =$ flow time). (After *Parker* and *Hedley* [88])

passing a minimum the relative drag increases again because the viscosity of the solution increases. The relative drag decreases significantly with increasing average molecular weight in the region where the influence of the solution viscosity is not yet critical. This region corresponds to about 1 to 10 ppm for the examples presented in Fig. 3.14. Fig. 3.15 shows how main-chain scission affects the relative drag with aqueous solutions of poly(acrylic acid) and polyacrylamide [88]. The solutions were forced repeatedly through a tube under constant conditions. It can be seen that the relative drag increased drastically with increasing number of passes. The effect was shown to be due to main-chain breakage. When asbestos fibers were subjected to the same treatment a similar behavior was observed, i.e. the relative drag increased with increasing number of passes when the suspension was subjected repeatedly to shear stress above a critical value.

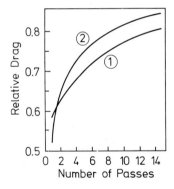

Fig. 3.15 The influence of repeated shearing on the drag reducing efficiency of poly-(acrylic acid) (1) at 2.5 ppm, MW $= 3.9 \times 10^7$ and polyacrylamide (2) at 5 ppm, MW $= 4.4 \times 10^6$ in aqueous solution. Driving pressure: 0.348 MN/m². Relative drag $= (t_{solution}/t_{water})^2$. $t =$ flow time. (After *Parker* and *Hedley* [88])

In conclusion, it can be stated, that a variety of polymers is capable of acting as drag reducing agents. Problems arise, however, as a consequence of main-chain scission, and so care must be taken in cases of prolonged or repeated utilization of polymers as drag reducers.

An analogous situation is encountered in the application of viscosity index improvers for lubricating oils in automobile engines. Polymers (commonly polyisobutene and polymethacrylates) are dissolved in the oil. On increasing the temperature, the fluidity of the oil increases, but the simultaneous expansion of the polymer coils strongly diminishes the extent of decrease of the macroviscosity. This is the basis for multigrade lubricating oil formulations. During application, lubricating oils are subjected to high shear rates (in the engine by the motion of the piston in the cylinder and in the pumping system). In order to preserve their thickening power lubricating oils have to exhibit, therefore, an appreciable shear stability. Since polymers have for many years been utilized as viscosity index improvers, it appears that this problem has been satisfactorily solved, although little information is available. This is certainly due to the fact that for proprietary reasons information concerning the composition of lubricating oil formulations is not freely disclosed.

For further details concerning viscosity index improvers the reader is referred to relevant publications [89—93].

3.4.3   Freezing and Thawing

Main-chain degradation induced by freezing polymer solutions has been occasionally investigated during the past 30 years [94—99]. The freezing stability of polymers becomes important in special situations where polymer solutions are cooled down to temperatures below the melting point. Frequently, for instance, polymer solutions are deaerated during laboratory experiments by freeze, pump and thaw cycles.

Main-chain scission induced by freezing may also cause problems when polymer solutions are transported in winter time.

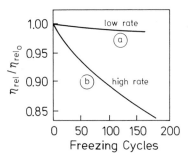

Fig. 3.16   Main-chain degradation of polyisobutene in the presence of O_2 induced by freezing. The relative viscosity vs. the number of freezing cycles. $\overline{M}_{v,0} = 2 \times 10^6$. Solvent: cyclohexane. Polymer concentration: 2 g/l.
(a) low freezing rate, (b) high freezing rate (After *Patat* and *Högner* [97]).

Generally, the density of a system changes during the transition from the liquid to the solid state. At the time of phase transition, stresses are generated which might induce in macromolecules strains strong enough for bond rupture. The extent of main-chain scission is dependent on the rate of cooling, the molecular weight, the polymer concentration, the concentration of dissolved gas and the solvent quality.

Relevant investigations were usually based on a series of many freezing cycles. It appears that at MW $< 10^6$ main-chain scission becomes noticeable only if the number of freezing cycles exceeds 50. Typical results, obtained by Patat and Högner [97], are shown in Fig. 3.16. It can be seen that at low freezing rates the relative viscosity of the poly-isobutene solution in cyclohexane was only slightly affected, whereas at high freezing rates a significant decrease in the relative viscosity was observed. In addition, it was concluded during that work that the extent of main-chain scission increases with polymer concentration, with improvement of solvent quality and with MW. More recently Åbbas and Porter [99] reported results obtained with polystyrene dissolved in p-xylene. For a sample with $\overline{M}_{w,0} = 1.13\ \overline{M}_{n,0} = 7.3 \times 10^6$, appreciable main-chain scission was already observed after a few freezing cycles. Changes in the MWD, as measured by GPC, can be seen in Fig. 3.17. From the results obtained at various polymer concentrations, it was concluded that above a transition range (1.0 to 2.5 g/l) main-chain rupture

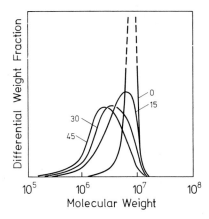

Fig. 3.17 Main-chain degradation of polystyrene in p-xylene solution induced by freezing. MWD as a function of the number of freezing cycles, as indicated at the curves. $\overline{M}_{w,0} = 7.3 \times 10^6$. Polymer concentration: 2.5 g/l. (After *Åbbas* and *Porter* [99])

occurred at random, whereas in highly dilute solutions another mechanism became operative. This conclusion follows from Fig. 3.18, where $\overline{M}_w/\overline{M}_{w,0}$ is plotted vs. $\overline{M}_n/\overline{M}_{n,0}$*). The solid line was calculated for random main-chain scission. The data for solutions with concentrations greater than 2.5 g/l are in agreement with the calculated curve, whereas the data obtained at lower concentrations deviate appreciably.

It can be concluded that in practice, main-chain degradation, induced by freezing of polymer solutions, can be ignored except when the MW is very high or when the number of freezing cycles significantly exceeds 10 to 50.

---

*) Refer to Section 1.4.1 and recall that $mu_1 = \overline{M}_n$ and $mu_2 = \overline{M}_w$. m designates the molecular weight of the base unit.

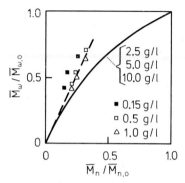

Fig. 3.18   Main-chain degradation of polystyrene in p-xylene solution induced by freezing. Relative change in weight average MW vs. relative change in number average MW at various polymer concentrations (as indicated on the graph). The solid line was calculated for random main-chain scission with $(\overline{M}_{w,0}/\overline{M}_{n,0}) = 1.13$. (After *Åbbas* and *Porter* [99])

3.4.4   Mastication of Natural and Synthetic Rubber

Mechanical degradation has been utilized for processing natural rubber since the middle of the 19th century [100]. The mechanical treatment of rubbers is called "mastication", a term derived from the Latin word "masticare" = to chew. Mastication is usually carried out on roll mills or in plasticators (internal mixers). In the presence of air it leads to a marked decrease in the average molecular weight of the rubber thus increasing its plasticity. Compounding ingredients can easily be added to the masticated rubber prior to vulcanization (crosslinking). Typical results demonstrating the decrease in the molecular weight as a function of the mastication time are shown in Fig. 3.19.

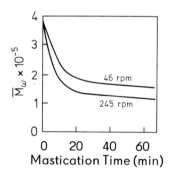

Fig. 3.19   Mastication of natural rubber in an internal mixer in the presence of air at ca. 52 °C and two different rotor speeds as indicated on the graph. (After *Harmon* and *Jacobs* [101])

Natural rubber (from Hevea brasiliensis) is composed of cis-1,4-polyisoprene. Under stress, homolytic bond scission is assumed to occur in the main chains leading to allyl type radicals:

$$\underset{\begin{subarray}{c}\\\end{subarray}}{\sim\overset{\overset{\textstyle CH_3}{|}}{C}=CH-CH_2-CH_2-\overset{\overset{\textstyle CH_2}{|}}{C}=CH\sim}\;\rightarrow\;\sim\overset{\overset{\textstyle CH_3}{|}}{C}=CH-CH_2^{\bullet}\;+\;{}^{\bullet}CH_2-\overset{\overset{\textstyle CH_3}{|}}{C}=CH\sim \qquad (3.14)$$

In an inert atmosphere mechanical treatment readily leads to intermolecular crosslinking as indicated by an increase of the average molecular weight and at a later stage by the formation of an insoluble and infusible threedimensional network. Typical results showing the increase in the average molecular weight as a function of mastication time are presented in Fig. 3.20 [102]. Branching and crosslinking become feasible through the reaction of macroradicals, formed according to reaction (3.14), with double bonds of other macromolecules

$$\underset{\underset{\overset{\overset{\textstyle CH_2}{|}}{\overset{\textstyle }{\S}}}{\overset{\textstyle }{\underset{\textstyle }{}}}}{\sim\overset{\overset{\textstyle CH_3}{|}}{C}=CH-CH_2^{\bullet}}\;+\;\underset{\underset{\overset{\overset{\textstyle CH_2}{|}}{\overset{\textstyle }{\S}}}{\overset{\textstyle CH}{||}}}{\overset{\overset{\textstyle CH_2}{|}}{C}-CH_3}\;\rightarrow\;\underset{\underset{\overset{\overset{\textstyle CH_2}{|}}{\overset{\textstyle }{\S}}}{\overset{\textstyle HC\cdot}{|}}}{\sim\overset{\overset{\textstyle CH_3}{|}}{C}=CH-CH_2-\overset{\overset{\textstyle CH_2}{|}}{C}-CH_3} \qquad (3.15)$$

Mastication aims at lowering the molecular weight and must, therefore, always be carried out under conditions which prevent crosslinking, i.e. in the presence of radical scavengers. Usually, molecular oxygen contained in the air is sufficient to completely suppress gel formation.

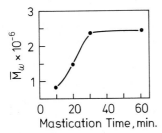

Fig. 3.20 Mastication of natural rubber in an argon atmosphere. Weight average molecular weight as a function of mastication time. (After *Dogadkin* and *Kuleznev* [102])

O_2 reacts with macroradicals to form peroxyl radicals which abstract hydrogen atoms intra- or intermolecularly from rubber (RH) or from additive molecules (AH), e.g.:

$$-CH_2-\overset{\overset{\textstyle CH_3}{|}}{C}=CH-CH_2^{\bullet}\;+\;O_2\;\rightarrow\;-CH_2-\overset{\overset{\textstyle CH_3}{|}}{C}=CH-CH_2-O-O\cdot \qquad (3.16)$$

$$-CH_2-\overset{\overset{\textstyle CH_3}{|}}{C}=CH-CH_2-O-O\cdot\;+\;RH\;\longrightarrow\;-CH_2-\overset{\overset{\textstyle CH_3}{|}}{C}=CH-CH_2-OOH\;+\;R\cdot \qquad (3.17)$$

(or AH)   (or A·)

Radicals R· and A· might react with each other, with O_2 or with peroxyl radicals. Alternatively they might react with double bonds in rubber molecules, giving rise to branching and crosslinking.

Mastication is usually carried out in the presence of O_2 at elevated temperatures *). Therefore, it is worthy to note, that the mastication efficiency decreases with increasing temperature due to the increased mobility of the macromolecules, which favors harmless energy dissipation. However, after passing a minimum it increases again. A typical efficiency vs. temperature curve is shown in Fig. 3.21, where the degree of degradation, α, achieved after a constant mastication time of 30 min, is plotted against the temperature. The increase of α at elevated temperatures is due to a thermo-oxidative reaction having an onset temperature of about 100 °C in the case of natural rubber.

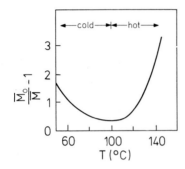

Fig. 3.21   Mastication of natural rubber. Degree of degradation $\alpha \equiv \dfrac{\overline{M}_0}{M} - 1$
(number of scissions per initial macromolecule) vs. temperature. Mastication time: 30 min. (From *Pike* and *Watson* [104] according to *Busse* [103])

It was found that mastication at elevated temperatures causes a higher extent of main-chain rupture than static oxidation [105]. Therefore, it can be concluded that oxidative main-chain scissioning is shear-induced, i.e. a mechanochemical initiation is followed by thermal oxidation, which is probably proceeding by a chain mechanism, as outlined in Chapter 1.

It is possible to accelerate the thermo-oxidative degradation with the aid of additives [106, 107], usually zinc salts or disulfides of aromatic or heterocyclic mercaptans (typical examples: zinc salt of pentachlorothiophenol and o,o′-dibenzamidophenyldisulfide). In addition, activators are used (e.g. tetraazoporphine or phthalocyanine-metal complexes (with Co, Cu, Fe)). In the presence of accelerators the minimum of the U-shaped efficiency-temperature curve is shifted to lower temperatures.

GPC analysis of natural rubber samples masticated at ca. 52 °C in the presence of air has revealed that the MWD is narrowed under the influence of shear. Typical results are presented in Fig. 3.22 [101]. It is interesting to note that rubber samples masticated for relatively short times possess a bimodal MWD with a small peak corresponding to

---

*) a temperature increase is frequently caused by mastication itself.

a molecular weight markedly higher than that of the starting material. This peak indicates the formation of crosslinks between the rubber molecules. With increasing mastication time, the high molecular weight peak disappears, the MWD becomes unimodal and narrower. This can be interpreted in terms of the fact that large macromolecules are broken preferentially. From various investigations it became evident that mastication-induced main-chain degradation with simultaneous end-group radical stabilization can only proceed down to a limiting molecular weight, estimated to be about 7×10^4 [100].

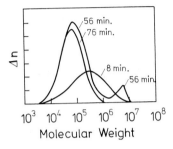

Fig. 3.22 Mastication of natural rubber. GPC analysis of samples masticated at $T \approx 52\,°C$ in the presence of air for various periods as indicated on the graph. (After *Harmon* and *Jacobs* [101])

The technique of mastication is also applicable to synthetic rubbers, such as styrene, nitrile and acrylic rubbers, polyisoprenes, polychloroprenes, butyl rubber and EPDM (a terpolymer of ethylene-propylene-diene monomers) [100]. Mastication is, however, of minor importance for the processing of synthetic rubbers, because their molecular weight can be regulated during polymerization.

3.4.5   Mechanochemical Synthesis of Block and Graft Copolymers

Inspired by the fact that macroradicals can be generated when polymers are subjected to mechanical treatment, polymer chemists have been searching for applications related to the synthesis of new polymers or to the modification of existing polymers.

As outlined in other chapters of this book, macroradicals can also be produced by irradiating macromolecules with UV light or high energy radiation, or by thermal treatment. A comparison with these methods reveals that mechanical generation of macroradicals has the advantage of producing primarily terminal radicals, making it most appropriate for the synthesis of block copolymers, a group of copolymers consisting of unbranched chains with long blocks*) of chemically identical base units arranged in series, e.g.:

$$A–A–A–A–A–A–A–A–A–A–A–A–B–B–B–B–B–B–B–B–B–B–B–B–B–B$$

However, it is frequently difficult to produce "pure" block copolymers. Quite often lateral macroradicals are produced during secondary reactions leading to the formation of graft copolymers, e.g.:

---

*) in the simplest case two blocks

$$B–B–B–B–B–B–B–B–B–B \sim$$
$$|$$
$$\sim\!A–A–A–A–A–A–A–A–A–A–A–A–A–A–A–A–A–A–A–A–A–A–A–A–A–A–A–A \sim$$
$$|$$
$$B–B–B–B–B–B–B–B \sim$$

Principally, two reaction modes can be employed for the production of block and graft copolymers, independent of the mode of free radical generation:

    (a) reactions in polymer–polymer systems
    (b) reactions in polymer–monomer systems

In mode (a), two or more different polymers are blended and subjected to mechanical treatment. With a blend of two polymers, block copolymers can be formed as a consequence of the following main-chain scission processes:

$$A_{m+n} \to A_m^\bullet + A_n^\bullet \tag{3.18}$$

$$B_{p+q} \to B_p^\bullet + B_q^\bullet \tag{3.19}$$

The combination of the macroradicals thus formed leads to the products:

$$A_m - B_p, \quad A_m - B_q, \quad A_n - B_p \quad \text{and} \quad A_n - B_q.$$

Frequently, disproportionation and homo-combination compete with hetero-combination, thus diminishing the block copolymer yield.

In mode (b), polymers are subjected to stress in the presence of one or more monomers with the result that the interaction of macroradicals with each other is inhibited by the reactions of monomer molecules with macroradicals, e.g.:

$$A_m^\bullet + pB \to A_m - B_p^\bullet \tag{3.20}$$

In both polymer–polymer and polymer–monomer systems, hydrogen abstraction reactions might occur, e.g.:

$$A_m^\bullet + PH \to A_mH + P\bullet \tag{3.21}$$

Here, PH denotes either an initial polymer molecule or a copolymer molecule just formed. Since the free radical site in $P\bullet$ (formed according to reaction (3.17)) is located laterally at the chain, reactions of $P\bullet$ give rise to the formation of graft copolymers, e.g. in the case of a polymer–polymer system:

$$\underset{\bullet}{\sim\!A\!\sim} + B_p^\bullet \to \sim\!A\!\sim \tag{3.22}$$
$$|$$
$$B_p$$

or in the case of a polymer–monomer system:

$$\underset{\bullet}{\sim\!A\!\sim} + pB \to \sim\!A\!\sim \tag{3.23}$$
$$|$$
$$B_p^\bullet$$

The importance of these grafting processes becomes obvious if we take into account that they are quite generally applicable to the great variety of natural and synthetic linear polymers. As far as polymer–polymer and polymer–monomer systems are concerned the number of possible combinations appears to be enormously large. Numerous publications have been devoted to mechanochemical syntheses [11, 25, 26, 108–112] and the amenability of mechanochemical processes to commercial applications has been demonstrated.

The importance of block and graft copolymers derives from the fact that, according to their composition, these copolymers can possess properties which are otherwise unattainable, e.g., the combination of different properties in one molecule for instance, hydrophilicity and hydrophobicity. For further information the reader is referred to other books [14, 113].

For practical applications, solid state treatment (commonly vibromilling) seems to be less suitable because of difficulties connected with grinding and comminution. Intimate mixing of the components is prerequisite for achieving high yields of graft or block copolymers. Therefore, processing in the molten state (e.g. coextrusion of two polymers) or in the viscoelastic (rubbery) state, appears to be most appropriate in order to synthesize new polymeric products under economically favorable conditions. In this respect mastication certainly has great advantages and numerous possibilities have been tested in order to modify elastomers and plastics [108, 114–118]. The endproducts of mechanically treated polymer–polymer and polymer–monomer systems are rather heterogeneous from the chemical point of view. They consist of a mixture of block, graft and homopolymers. According to a proposal by *Baramboim* the term "interpolymers" should be used for these composite materials.

It should be pointed out that block and graft copolymers can also be synthesized by subjecting solutions of polymer–polymer or polymer–monomer systems to high-speed stirring or ultrasonic irradiation. These methods, especially ultrasonic treatment, allow easy production of block copolymers on a laboratory scale [40, 77, 78, 119–122].

References to Chapter 3

[1] *A. Casale* and *R. S. Porter*, "Polymer Stress Reactions", Vol. 1 and 2, Academic Press (1978), New York.
[2] *H.-H. Kausch*, "Polymere Fracture", Vol. 2 of "Polymers, Properties and Applications", Springer-Verlag, Berlin-Heidelberg-New York (1978).
[3] *E. H. Andrews* and *P. E. Reed*, "Molecular Fracture in Polymers", Adv. Polym. Sci. 27, 1 (1978).
[4] *J. G. Williams*, "Application of Linear Fracture Mechanics", Adv. Polym. Sci. 27, 67 (1978).
[5] *C. B. Bucknall*, "Fracture and Failure of Multiphase Polymers", Adv. Polym. Sci. 27, 121 (1978).
[6] *A. M. Basedow* and *K. Ebert*, "Ultrasonic Degradation of Polymers in Solution", Adv. Polym. Sci. 22, 83 (1977).
[7] *K. Murakami*, "Mechanical Degradation" in *H. H. G. Jellinek* (ed.), "Aspects of Degradation and Stabilization of Polymers", Elsevier, Amsterdam (1978), p. 295.
[8] *J. P. Sohma* and *M. Sakaguchi*, "E.S.R.Studies on Polymer Radicals Produced by Mechanical Destruction and Their Reactivity", Adv. Polym. Sci. 20, 111 (1976).
[9] *B. Rånby* and *J. Rabek*, "ESR Spectroscopy in Polymer Research", Springer-Verlag, Heidelberg (1977).
[10] *H. Fischer*, "Magnetische Eigenschaften freier Radikale" in *K.-H.* and *A. M. Hellwege* (eds.), Landolt-Börnstein, Neue Serie II/1, Springer-Verlag, Berlin (1965).
[11] *N. K. Baramboin*, "Mechanochemistry of Polymers", Maclaren, London (1964).

[12] *W. F. Watson*, "Mechanochemical Reactions" in *E. M. Fettes* (ed.), "Chemical Reactions of Polymers", Wiley, New York (1964).

[13] *D. C. Allport*, "Mechanochemical Syntheses" in *D. C. Allport* and *W. H. Janes* (eds.), Block Copolymers, Applied Science Publ., London (1973).

[14] *R. J. Ceresa*, "Block and Graft Copolymerization", Wiley, New York (1973).

[15] *H. H. Kausch*, J. Macromol. Sci., Macromol. Chem., C 4, 243 (1970).

[16] *S. Bresler*, *A. Osminskaya* and *A. Popov*, Kolloid-Z. 20, 403 (1958), Zh. Techn. Fiz. 29, 358 (1959).

[17] *P. Y. Butyagin* and *A. Berlin*, Vysokomol. Soed. 1, 865 (1959).

[18] *P. Y. Butyagin*, *I. V. Kolbanev* and *V. A. Radtsig*, Sov. Phys. Sol. State 5, 1642 (1964).

[19] *S. N. Zhurkov*, *A. Y. Savostin* and *E. E. Tomashevskii*, Dokl. Akad. Nauk USSR, 159, 303 (1964).

[20] *O. P. Campbell* and *A. Peterlin*, J. Polym. Sci. B 6, 481 (1968).

[21] *A. Peterlin*, J. Polym. Sci. C 32, 297 (1970).

[22] *A. Peterlin*, "ESR Application to Polymer Research" (Nobel Symp. 22, *Kinell* and *Rånby*, eds.), Almqvist and Wilsell, Stockholm (1973).

[23] *K. L. DeVries*, J. Polym. Sci. C 32, 325 (1970).

[24] *K. B. Åbbas* and *R. S. Porter*, J. Appl. Polym. Sci. 20, 1289 (1976).

[25] *W. Lauer*, Kautschuk, Gummi, Kunststoffe 28, 536 (1975); 28, 608 (1975).

[26] *N. K. Baramboim* and *W. G. Protasow*, Technik 30, 73 (1975).

[27] *H. Deuel* and *R. Gentili*, Helv. Chim. Acta 39, 1586 (1956).

[28] *R. E. Benson* and *J. E. Castle*, J. Phys. Chem. 62, 840 (1958).

[29] *N. A. Plate* and *V. A. Kargin*, J. Polym. Sci. C 4, 1027 (1963).

[30] *H. Grohn* and *R. Paudert*, Z. Chem. 3, 89 (1963).

[31] *H. Grohn*, *R. Paudert* and *B. Hosselbarth*, Plaste Kautsch. 13, 1 (1966).

[32] *A. A. Berlin* and *A. M. Dubinskaya*, Vysokomol. Soedin. 1, 1678 (1959).

[33] *B. E. Geller*, *G. V. Goryachko*, *N. A. Dimitriera* and *N. I. Larionov*, Vysokomol. Soedin. 1, 1610 (1959).

[34] *N. K. Baramboim* and *Yu. S. Simakov*, Polym. Sci. USSR 8, 255 (1966).

[35] *A. Guyot* and *A. Michel*, J. Appl. Polym. Sci. 13, 911 (1969).

[36] see chapter 5 of Ref. [2].

[37] *Ya. I. Frenkel*, Acta Physicochim. USSR 19, 51 (1944).

[38] *W. J. Kauzmann* and *H. Eyring*, J. Am. Chem. Soc. 62, 3113 (1940).

[39] *F. Bueche*, "Physical Properties of Polymers", Wiley, New York (1962).

[40] *H. Fujiwara*, *K. Okazaki* and *K. Goto*, J. Polym. Sci., Polym. Phys. Ed. 13, 953 (1975).

[41] *R. J. Nash* and *D. M. Jacobs*, Faraday Disc., Solid State Interfaces (1972).

[42] *R. S. Porter*, *M. R. J. Cantow* and *J. F. Johnson*, J. Appl. Polym. Sci. 11, 335 (1967).

[43] see Ref. [2], Chapter 8.

[44] *S. N. Zhurkov*, *V. A. Zakrevskii*, *V. E. Korsukov* and *V. S. Kuksenko*, Fizika Tverdogo Tela 13, 2004 (1971); Soviet Phys. Solid State 13, 1680 (1972); J. Polymer Sci. A-2, 10, 1509 (1972).

[45] *S. N. Zhurkov* and *V. E. Korsukov*, Fizika Tverdogo Tela 15, 2071 (1973); Soviet Phys. Solid State 15, 1379 (1974).

[46] *V. A. Zakrevskii* and *V. E. Korsukov*, Vysomol. Soed. B 13, 105 (1971); A 14, 955 (1972).

[47] *A. Peterlin*, Intern. J. Fracture 11, 761 (1975).

[48] *T. M. Stoeckel*, *J. Blasius* and *B. Crist*, J. Polym. Sci. Polym. Phys. Ed. 16, 485 (1978).

[49] *A. Ram* and *A. Kadim*, J. Appl. Polym. Sci. 14, 2145 (1970).

[50] *F. Krause* and *H. Grohn*, Plaste Kautsch. 11, 6 (1964).

[51] *C. Popa*, cited by *C. Simionescu* and *C. Vasiliu-Oprea*, "Mechanochimia Computilor Macromoleculari", Acad. Rep. Soc., Romania, Bucaresti (1967).

[52] *Y. Won*, *T. Fukutomi*, *T. Kakurai* and *T. Noguchi*, Kobunshi Kagaku 27, 594 (1970).

[53] *T. Fukutomi*, *M. Tsukada*, *T. Kakurai* and *T. Noguchi*, Polymer J. 3, 717 (1972).

[54] *R. Simha*, J. Appl. Phys. 12, 569 (1941).

[55] *K. Arai, K. Nakamura, T. Komatsu* and *T. Nakagawa*, Kogyo Kogaku Zashi 71, 1438 (1968).
[56] *H. Grohn* and *F. Krause*, Plaste Kautsch. 11, 2 (1964).
[57] *J. W. Breitenbach, J. K. Rigler, B. A. Wolf*, Makromol. Chem. 164, 353 (1973).
[58] *A. B. Bestul* and *H. V. Belcher*, Phys. Rev. 79, 223 (1950).
[59] *P. Goodman*, J. Polym. Sci. 25, 325 (1954).
[60] *R. S. Porter* and *J. F. Johnson*, J. Phys. Chem. 63, 202 (1959); J. Appl. Phys. 35, 3149 (1964).
[61] *R. S. Porter, M. J. R. Cantow* and *J. F. Johnson*, J. Polym. Sci. C 16, 1 (1967); Polymer 8, 87 (1967).
[62] *P. J. Sheth* and *J. F. Johnson*, "Ultrasonic Irradiation" in *A. Casale* and *R. S. Porter*, "Polymer Stress Reactions" (see [1]).
[63] *L. Bergmann*, "Der Ultraschall", Hirzel-Verlag, Stuttgart (1957).
[64] *W. Schaaffs*, "Molekularakustik", Springer-Verlag, Berlin (1963).
[65] For relevant references refer to [6].
[66] *A. Weissler*, J. Appl. Phys. 21, 171 (1950).
[67] *B. E. Noltingk* and *E. A. Neppiras*, Proc. Phys. Soc. B 63, 674 (1950).
[68] *W. Gueth*, Acustica 6, 526 (1956).
[69] *W. Gueth* and *E. Mundry*, Acustica 7, 241 (1957).
[70] *J. Schmid*, Acustica 9, 321 (1959).
[71] *G. Gooberman*, J. Polym. Sci. 42, 25 (1960); 47, 229 (1960).
[72] *G. Gooberman* and *J. Lamb*, J. Polym. Sci. 42, 35 (1960).
[73] *M. Okuyama* and *N. Sata*, Z. Elektrochem. 58, 197 (1954).
[74] *M. Okuyama* and *T. Hirose*, J. Appl. Polym. Sci. 7, 591 (1963).
[75] *N. Suppanz*, Thesis, Universität Heidelberg (1972).
[76] *J. R. Thomas*, J. Phys. Chem. 63, 1725 (1959).
[77] *A. Henglein*, Makromol. Chem. 14, 15 (1954); 15, 188 (1955).
[78] *A. Henglein*, Makromol. Chem. 18, 37 (1956).
[79] *H. Fujiwara, K. Okazaki* and *K. Goto*, J. Polym. Sci. 13, 953 (1975).
[80] *A. M. Basedow* and *K. H. Ebert*, Makromol. Chem. 176, 745 (1975).
[81] *P. J. Sheth, J. F. Johnson, R. S. Porter*, Polymer 18, 741 (1977).
[82] *B. M. E. Van der Hoff* and *P. A. R. Glynn*, J. Macromol. Sci. Chem. A 8, 429 (1974).
[83] *B. A. Toms*, Proc. 1st Intern. Congr. Rheology, Vol. II, p. 135 (1949).
[84] *P. R. Kenis*, J. Appl. Polym. Sci. 15, 607 (1971).
[85] *T. Hasegawa* and *Y. Tomita*, Bull. JSME 17, 73 (1974).
[86] *H. W. Friebe*, Rheol. Acta 15, 329 (1976).
[87] *R. Y. Ting* and *D. L. Hunston*, Ind. Eng. Chem., Proc. Res. Dev. 16, 129 (1977).
[88] *C. A. Parker* and *A. H. Hedley*, J. Appl. Polym. Sci. 18, 3403 (1974).
[89] *C. Lindhe*, J. Chromatogr. Sci. 9, 420 (1971).
[90] *J. Jakobsen, D. M. Sanborn* and *W. O. Winder*, SAE Techn. Paper 730482, p. 59 (1973).
[91] *E. E. Klaus* and *M. R. Fenske*, Lubr. Eng. 11, 101 (1955).
[92] *T. W. Johnson* and *M. T. O'Shaughness*, SAE/ASTM Symp. Viscosity and Lubrication, Detroit, Mich. (1977).
[93] Ref. [1], Vol. 2, p. 548—551, and references therein.
[94] *N. I. Nikitin* and *P. I. Klenkova*, Issled. Obl. Vysokomol. Soed. 138 (1949).
[95] *S. Oprea* and *Cr. Simionescu*, Vysokomol. Soed. 8, 1132 (1966).
[96] *A. A. Berlin, E. A. Penskaya* and *G. I. Volkova*, J. Polym. Sci. 56, 477 (1962).
[97] *F. Patat* and *W. Högner*, Makromol. Chem. 75, 85 (1964).
[98] *H. H. G. Jellinek* and *S. Y. Fok*, Makromol. Chem. 104, 18 (1967).
[99] *K. B. Åbbas* and *R. S. Porter*, J. Polym. Sci., Polym. Chem. Ed. 14, 553 (1976).
[100] for relevant references see Ref. [1] Vol. 2, p. 449—465.
[101] *D. J. Harmon* and *H. L. Jacobs*, J. Appl. Polym. Sci. 10, 253 (1966).
[102] *B. A. Dogadkin* and *V. N. Kuleznev*, Kolloid Zh. 20, 674 (1958).

[103] *W. F. Busse*, Proc. Rubber Techn. Conf. 288 (1938).
[104] *M. Pike* and *W. Watson*, J. Polym. Sci. 9, 229 (1952).
[105] *L. Mullins* and *W. F. Watson*, J. Appl. Polym. Sci. 1, 245 (1959).
[106] Ullmanns Encyklopädie der technischen Chemie. 4th ed., Vol. 13, p. 637, Verlag Chemie, Weinheim (1977).
[107] *S. Koch* (ed.), Bayer-Handbuch für die Gummi-Industrie, Bayer AG, Leverkusen (1971).
[108] *A. Casale* and *R. S. Porter*, Adv. Polym. Sci. 17, 1 (1975); ref. [1] Vol. 1, p. 161—252.
[109] *M. K. Akutin*, Plast. Inst. Trans. 28, 216 (1960).
[110] *M. L. Kerber*, Sov. Plast. 5, 64 (1971).
[111] *R. J. Ceresa*, J. Polym. Sci. 53, 9 (1961).
[112] *C. Vasiliu-Oprea* and *Cr. Simionescu*, Mater. Plast. 3, 64 (1966).
[113] *W. J. Burlant* and *A. S. Hoffman*, "Block and Graft Copolymers", Van Nostrand-Reinhold, Princeton, N.J. (1960).
[114] *D. J. Angier* and *W. F. Watson*, J. Polym. Sci. 18, 129 (1955); 20, 235 (1956) and 25, 1 (1957); Brit. Pat. 828, 895 (1960).
[115] *R. J. Ceresa*, *D. J. Elliott* and *W. F. Watson*, Brit. Pat. 851, 731 (1960).
[116] *A. A. Berlin*, Usp. Khim. 27, 94 (1958).
[117] *R. J. Ceresa* and *W. F. Watson*, Trans. Inst. Rubber Ind. 35, 19 (1959).
[118] *J. LeBras* and *P. Compagnon*, Bull. Soc. Chim. 11, 553 (1944).
[119] *G. Schmid* and *A. Henglein*, Kolloid-Z. 148, 73 (1956).
[120] *A. Nakamo* and *Y. Minoura*, J. Appl. Polym. Sci. 15, 927 (1971).
[121] *A. A. Berlin* and *A. M. Dubinskaja*, Vysokomol. Soedin 2, 1426 (1960).
[122] *Z. Osawa*, *T. Kimura* and *I. Kasuga*, J. Polym. Sci. 7, 2007 (1969).

4   Photodegradation

4.1   Introduction

4.1.1   General Remarks

Most commercial organic polymers undergo chemical reactions upon irradiation with ultraviolet (UV) light, because they possess chromophoric groups (as regular constituents or as impurities) capable of absorbing UV light. This fact is important because the spectrum of the sunlight penetrating the earth's atmosphere contains a portion of UV light. Therefore, photoreactions are usually induced when organic polymers are subjected to outdoor exposures. In general, photoreactions in commercial polymers are harmful: they cause embrittlement and color changes. Ever since large scale production of polymers started a few decades ago, the plastics producer's interest lay in developing methods to prevent photodegradation. In this respect, photolytical reactions of special importance are UV light initiated oxidative chain reactions, i.e. autoxidative processes (see Chapter 1). Thus, in this chapter the photooxidation of polymers will be discussed in some detail, after a concise treatment of some elementary facts such as light absorption, modes of photochemical reactions and mechanistic aspects.

Apart from deterioration effects, the field of photodegradation comprises beneficial aspects also. A typical example is the utilization of readily degradable polymers as positively acting resists for the production of solid state electronic microstructures (e.g. integrated circuits).

Another interesting development concerns, moreover, polymers with predictable lifetimes. In this case advantage is taken of the fact that certain low molecular weight compounds are capable of acting as photo-inhibitors, whereas their photolysis decomposition products accelerate the photodegradation of the polymer.

A series of new applications might develop in the future, since powerful lasers of various wavelengths are becoming commercially available and laser-induced chemical reactions are being investigated in various laboratories.

It is appropriate to emphasize here two important aspects:

(i) the specific interaction of light with organic compounds and
(ii) the randomness of photochemical reactions in polymers.

As will be explained in some detail below, light absorption in a molecule consists of a *specific interaction* of a certain chromophoric group with a photon of given energy. The remainder of the molecule remains unaffected during the absorption act. Knowledge about absorptivities of chromophores, is the synthetic polymer chemist's tool for attaining photochemical selectivity. In other words, if a polymer chain is supposed to be ruptured at a certain position on irradiation, this goal can be achieved by synthetically introducing an appropriate chromophore at that place in the polymer backbone.

The *second aspect*, referred to above, derives from the fact that light is absorbed statistically by the chromophores in a system. All that we know, is the probability for the absorption of a certain photon by a certain chromophore. We do not know, however, when this will happen. With regard to a light absorbing homopolymer, this implies equal probability for the absorption of a photon by all base units. Chemical reactions occurring subsequent to light absorption can be initiated at any place in the macro-molecule. Implications of the random occurrence of chemical reactions in polymers have been discussed earlier in Section 1.3.2.

Owing to its enormous practical importance, photodegradation, in its many aspects is intensely studied in a great number of places around the globe and a very large body of papers is published every year. Here, the reader's attention is drawn to several books and articles recommended for further reading [1—16].

4.1.2 Light Sources

As far as practical applications of polymers are concerned, the sun is the most important light source. The spectrum of sunlight penetrating to the earth's surface ranges from about 290 to 3000 nm. The spectral distribution depends on atmospherical conditions and the latitude. Somewhat less than 10% of the sunlight at the earth's surface is UV-light, about 50% is visible and about 40% is IR light. For laboratory and industrial irradiations various types of lamps are available. Emission spectra of several lamps are shown in Fig. 4.1. Frequently, mercury lamps are used: low pressure lamps with

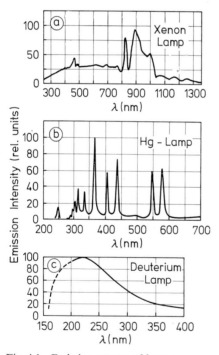

Fig. 4.1 Emission spectra of lamps.
(a) Xenon lamp, Osram, XBO 1 600 W; (b) High pressure mercury lamp, Philips, 1 000 W; (c) Deuterium lamp, Original Hanau D 200 F, 200 W

two intense lines at 184.9 and 253.8 nm and medium pressure lamps with a great number of lines, the most intense one corresponding to 366 nm. For preparative purposes, high pressure Hg lamps are most suitable because of their high emission intensities. Most frequently used are the lines at 254, 265, 313, 366, 436, and 546 nm. Carbon arc and xenon arc lamps are utilized quite often in devices for accelerated weathering tests etc. More details about these standard light sources can be found in ref. [7].

Recently powerful lasers have become commercially available. They emit coherent and monochromatic light. High power lasers are operated in a pulsed mode, in many cases, with adjustable pulse repetition rates. Preparative applications are becoming feasible. Appropriate literature for further information is available [17−19]. Operation of high power lasers in a single pulse mode is possible. Pulse durations of several nanoseconds are quite common at present and with special techniques, subnanosecond pulse lengths are attainable. Thus, with the aid of laser flash photolysis, it is now possible to study mechanisms and kinetics of many photochemical reactions that proceed very rapidly. Table 4.1 shows a compilation of typical lasers.

Table 4.1   Commercially available high power lasers [18, 19]

Type	λ (nm)	Pulse Energy (mJ)	Remarks *)
ruby	694	several 1 000	solid state, pulse repetition rate: 1 Hz,
	347	300	frequency doubled
Nd-YAG	1 060	several 1 000	solid state, pulse repetition rate: 20 Hz
	530	300	frequency doubled
	265	50	frequency quadrupled
N_2^+	337	9	gas
ArF	193	110	gas, excimer
KrF	249	500	gas, excimer
XeCl	308	250	gas, excimer
XF	351	180	gas, excimer
CO_2	10 600	2 800	gas

*) Single pulse operation is possible in all cases; the pulse repetition rate for the gas lasers is adjustable between 0.05 and 20 Hz; for further specifications see: Laser Focus, 1981 Buyer's Guide, Advanced Technology Publications, Newton, Mass.

4.1.3   Light Absorption and Quantum Yield

The absorption of light is prerequisite for the occurrence of photochemical reactions. Saturated compounds possessing bonds such as C−C, C−H, O−H, C−Cl etc. absorb light at $\lambda \leq 200$ nm. Carbonyl groups and conjugated C=C bonds absorb above $\lambda = 200$ nm and have absorption maxima between 200 and 300 nm. Fig. 4.2 shows absorption spectra of several polymers. It should be pointed out that only a small

7 Schnabel, Polymer Degr.

number of the important polymers are capable of absorbing solar radiation. However, quite frequently commercial polymers contain impurities capable of absorbing sunlight. This explains, in most cases, the instability of polymers which, according to their chemical constitution, should be resistant to solar radiation. The chance of an absorbed photon to induce a chemical change in the molecule depends principally on the photophysical processes following the absorption act.

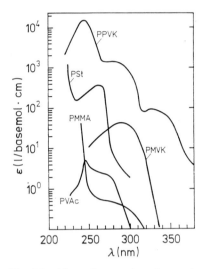

Fig. 4.2   Absorption spectra of several polymers recorded in dioxane solution. PMMA = polymethylmethacrylate, PVAc = polyvinylacetate, PSt = polystyrene, PMVK = poly(methylvinyl ketone), PPVK = poly(phenylvinyl ketone)

According to the energy state diagram in Fig. 4.3, the absorption of a photon can proceed either as $S_0 + h\nu \rightarrow S_1$ or as $S_0 + h\nu \rightarrow T_1$. The extinction coefficients differ appreciably: $\varepsilon(S_0 \rightarrow S_1) \gg \varepsilon(S_0 \rightarrow T_1)$. Therefore $S_0 \rightarrow T_1$ processes (designated as "forbidden") are negligible with respect to photochemical changes. Commonly, owing to their relatively long lifetimes, chemical reactions originate from S_1 or T_1 states. T_1 states are rather long-lived (k_P: $10^6 - 10^{-1}$ s^{-1}) owing to the fact that transitions of the type $T_1 \rightarrow S_0$ are "forbidden". Internal conversions are very rapid processes ($k_{IC} \approx 10^{10}$ s^{-1}).

Radiative and radiationless deactivation processes of S_1 states occur rapidly. Generally, fluorescence occurs much faster than phosphorescence, $k_F \gg k_P$. Scheme 4.1 presents important photophysical reactions. It is interesting to note that multiphotonic (mostly biphotonic) processes are becoming increasingly important for photodegradation applications, especially in connection with the use of lasers. Scheme 4.2 depicts two typical examples.

Usually, it is possible to determine the absorbed dose quite accurately, i.e. the number of photons absorbed per unit mass or unit volume by a sample during irradiation. Appropriate physical and chemical actinometers are available which can be easily handled.

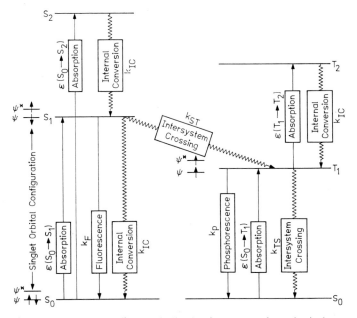

Fig. 4.3   Energy state diagram indicating important photophysical processes. Absorptive properties are characterized by extinction coefficients ε based on the relationship $I = I_0 \exp(-\varepsilon cd)$, where I_0 denotes the intensity of light incident upon a layer of optical path length d containing chromophores of concentration c. Transitions from excited energy states to states of lower energy (including the ground state) are charaterized by 1st order rate constants k (unit: time^{-1})

Scheme 4.1   Important photophysical processes. Superscripts 1, 2, and 3 denote singlet, doublet and triplet states; asterisks denote excited states

Photon	$^1M + h\nu \rightarrow {}^1M^*$ (1st excited singlet state)
Absorption	$^1M + h\nu \rightarrow {}^1M^{**}$ (higher excited singlet state)
	$^1M + h\nu \rightarrow {}^2M_{\bullet}^{+} + {}^2e^-$ (photoionization)
Radiationless	$^1M^* \rightarrow {}^1M +$ energy (internal conversion)
Transitions	$^1M^* \rightarrow {}^3M^* +$ energy (intersystem crossing)
	$^1M^{**} \rightarrow {}^1M^* +$ energy (internal conversion)
	$^3M^{**} \rightarrow {}^3M^* +$ energy (internal conversion)
Luminescence	$^1M^* \rightarrow {}^1M + h\nu$ (fluorescence)
(Radiative Transitions)	$^3M^* \rightarrow {}^1M + h\nu$ (phosphorescence)

Scheme 4.2   Multiphotonic processes (typical cases)

$$^1M \ + h\nu \ \rightarrow \ ^1M^* \rightarrow \ ^3M^*$$

$$^3M^* + h\nu' \ \rightarrow \ ^3M^{**}$$ $\Big\}$ excitation of higher triplet states

$$^1M \ + h\nu \ \rightarrow \ ^1M^*$$ "simultaneous" absorption

$$^1M^* + h\nu \ \rightarrow \ ^1M^{**}$$ of monoenergetic photons

$$^1M \ + nh\nu \rightarrow \ ^2M_{\cdot}^+ + \ ^2e^-$$ multiphotonic ionization

Frequently, photochemists use the unit Einstein = one mol of photons. The conversion is expressed as the "quantum yield", i.e. the number of atoms or molecules converted per photon absorbed by the irradiated material. In photodegradation of polymers, the quantum yield for main-chain scission, $\Phi(S)$, for example, denotes the number of main-chain scissions per photon absorbed by the polymer.

Further information on principles of photophysics and photochemistry is available from various books [20−26].

4.1.4   Chemical Reactions

Photochemical reactions can originate from radical ions or excited states. If we concentrate the discussion on irradiation with light of photon energies smaller than the ionization potential, chemical changes can be initiated either by exciting molecules to repulsive states or by generating long-lived excited states (usually T_1 states) which are capable of interacting chemically with other molecules. It was pointed out in the preceding section that radiative and radiationless i.e. physical deactivation routes are always competing with chemical routes. Quite seldom one observes quantum yields close to unity for chemical conversions. Usually, only a small portion of the total excited states deactivates via chemical routes. With respect to the initiation of photochemical reactions, it is appropriate to distinguish primary from secondary photochemical processes. In a primary process, the excited molecule dissociates into free radicals, for example:

$$M^* \ \rightarrow \ R_1^{\cdot} + \ R_2^{\cdot} \tag{4.1}$$

A typical example pertains to α-scission of ketones (also referred to as Norrish type I processes):

$$R_1 - \underset{\underset{O}{\|}}{C} - R_2 \ + \ h\nu \ \rightarrow \ R_1 - \underset{\underset{O}{\|}}{C} \cdot \ + \ R_2^{\cdot} \tag{4.2}$$

A typical secondary photochemical process is depicted in reaction (4.3):

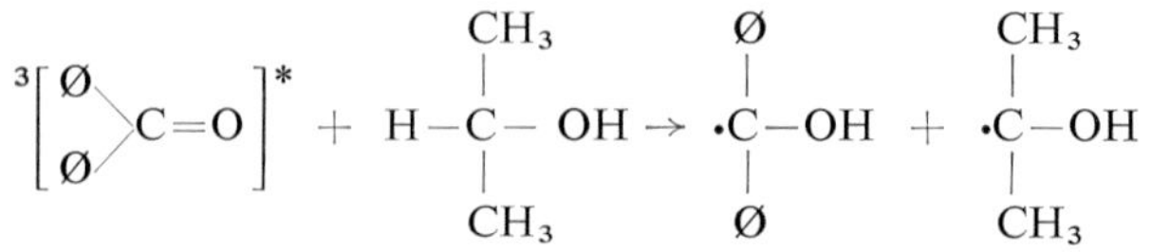

$$^3\left[\underset{\varnothing}{\overset{\varnothing}{>}} C = O \right]^* + \underset{\underset{CH_3}{|}}{\overset{\overset{CH_3}{|}}{H - C}} - OH \rightarrow \underset{\underset{\varnothing}{|}}{\overset{\overset{\varnothing}{|}}{\cdot C}} - OH + \underset{\underset{CH_3}{|}}{\overset{\overset{CH_3}{|}}{\cdot C}} - OH \tag{4.3}$$

In this case a rather long-lived triplet-excited ketone (benzophenone) reacts with a hydrogen donor (2-propanol) to form ketyl and 2-hydroxy-2-propyl radicals.

The generation of ions becomes feasible under certain conditions, e.g., if triplets $^3M_1^*$ form exciplexes E with ground state molecules of a second compound $^1M_2$ present in the system:

$$^3M_1^* + {^1M_2} \rightleftarrows E \quad \begin{array}{l} \longrightarrow R_1^{\cdot} + R_2^{\cdot} \\ \longrightarrow A^{(-)} + K^{(+)} \end{array}$$

(4.4a)
(4.4b)

Exciplexes are complexes of excited chromophores and non-excited molecules of different chemical nature. Ions are formed if exciplexes dissociate according to reaction (4.4b).

Moreover, ionic transients can be produced upon irradiation of charge transfer complexes with light of wave lengths corresponding to photon energies significantly smaller than the ionization energy of the constituents of the complex:

$$DA \overset{h\nu}{\underset{}{\rightleftarrows}} D_{\cdot}^+ + A_{\cdot}^- \tag{4.5}$$

According to the definition given above, reaction (4.5) should be classified as a primary process.

4.2 Mechanistic Aspects

4.2.1 Excited States, Free Radicals and Ionic Species

The initiation of photochemical processes can be understood in terms of the formation and reaction of electronically excited molecules, free radicals and ions, or radical ions, respectively, as was shown in the preceding section. Some characteristic cases will be discussed in order to show how the existence of transient species can be demonstrated. Molecules in *electronically excited states* can be detected either by luminescence or by light absorption. As has been pointed out in Section 4.1.3 there are two modes of luminescence: fluorescence and phosphorescence, corresponding to the transitions $S_1 \rightarrow S_0$ and $T_1 \rightarrow S_0$, respectively (see Fig. 4.3). Fluorescence spectra obtained with de-aerated CH_2Cl_2 solutions of polystyrene at room temperature are shown in Fig. 4.4 [27]. A gating technique was used in this case which allowed the recording of emission spectra at definite times after the excitation (duration: 10 ns). At the end of the flash, a spectrum essentially composed of the emission from "monomer" singlet states was observed (spectrum I in Fig. 4.4). The spectrum taken 45 ns after the flash is due to excimer singlet states. Excimers are complexes of excited and non-excited, (ground state) chemically identical chromophores, in this case phenyl groups, as shown for benzene and polystyrene in reactions (4.6a) and (4.6b), respectively

(4.6a)

ground state   "monomer" singlet   "excimer" singlet

(4.6b)

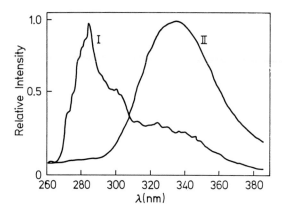

Fig. 4.4   Fluorescence spectra of polystyrene, obtained in CH$_2$Cl$_2$ solution (1 g/l) in the absence of O$_2$. λ_{exc} = 257 nm. Flash duration: ca. 10 ns. (After *Phillips* et al. [27]).

(I) recorded at the end of the flash, (II) recorded 45 ns after the flash

Analogous results were obtained with poly-1-vinylnaphthalene [27]. The decay lifetimes were determined as 1 ns (monomer) and 19 ns (excimer) for polystyrene and as 7 ns (monomer) and 43 ns (excimer) for poly-1-vinylnaphthalene. The excimers could also be detected by light absorption as shown in Fig. 4.5. The maximum at 530 nm (2.4 eV) corresponds to the transition $S_1 \rightarrow S_3$ [28].

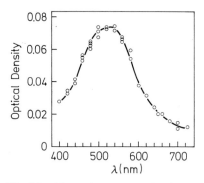

Fig. 4.5   Absorption spectra of polystyrene, obtained in p-dioxane solution (10^{-1} g/l) in the absence of O$_2$ at the end of a laser flash. λ_{exc} = 265 nm. Flash duration: 25 ns. (After *Tagawa* and *Schnabel* [28])

Excited triplet states can be detected by phosphorescence or by light absorption (triplet-triplet absorption via $T_1 \rightarrow T_n$ transitions, as depicted in Fig. 4.3). Fig. 4.6 presents the triplet-triplet spectra of polyvinylbenzophenone and p-isopropylbenzophenone (a low molecular weight model compound) [29]. The spectra of both compounds are quite similar. However, the decay rates differ appreciably as indicated by the oscilloscope traces, also shown in Fig. 4.6. The polymer effect is due to the high probability for the occurrence of triplet-triplet annihilation and intramolecular selfquenching processes, i.e. it is caused by the high local concentration of chromophores in the case of poly-vinylbenzophenone.

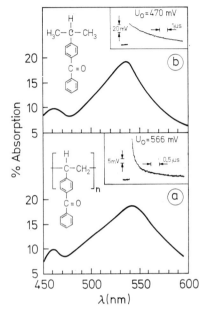

Fig. 4.6 Triplet-triplet absorption spectra of polyvinylbenzophenone (a) and p-iso-propylbenzophenone (b) obtained at the end of a 25 ns flash of 347 nm light in Ar-saturated benzene solution at room temperature. [Polymer]: 5.4×10^{-4} base mol/l; $\overline{M}_w$: 2×10^5. [model]: 7.4×10^{-4} mol/l. Insets: Oscilloscope traces depicting the decay of the absorption. (After *Schnabel* [29])

Luminescence techniques are highly appropriate for the detection of low concentrations of UV-light absorbing "impurities" in otherwise transparent polymers such as poly-ethylene and polypropylene [3]. These impurities can be incorporated in main chains (backbone-chromophores) or in side groups. On the other hand, impurities can be distri-buted more or less randomly without being chemically bound to the polymer matrix. Owing to their ability to absorb light, they can play an important role as initiators for photochemical degradation processes. Fig. 4.7 shows the fluorescence excitation spectra of polyethylene, polypropylene and poly-4-methylpentene-1. By comparison with the absorption spectrum of a typical aliphatic unsaturated ketone it becomes evident that the polymers contain impurities of analogous composition [3].

Frequently, the rate of photo-oxidative processes can be monitored by recording the change of luminescence intensity as a function of time. In polypropylene, for example, the unsaturated ketone was found to be gradually consumed during irradiation under simulated sunlight conditions [3].

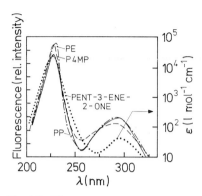

Fig. 4.7   Fluorescence excitation spectra of films of polyethylene (PE), polypropylene (PP) and poly-4-methylpentene-1 (P4MP). For comparison: absorption spectrum of pent-3-ene-2-one. (After *Allen* and *McKellar* [3])

Free radicals are almost ubiquitous in photochemical processes since homolytic bond dissociation is the usual pathway in *chemical* deactivation of excited states. The electron spin resonance (ESR) method allows the detection, and quite often the identification, of free radicals. Characteristic examples of free radicals generated during the photolysis of polymers are compiled in Table 4.2. This subject has been reviewed by *Tsuji* [30] and by *Rånby* and *Rabek* [31]. Free radicals are highly reactive. They readily combine or attack intact molecules. Therefore, it is often difficult to detect "primary" radicals, generated in bond homolysis. Frequently, "secondary" free radicals have been identified by ESR, i.e. radicals formed via a reaction of primary radicals with intact molecules, such as

$$R_1^{\cdot} + R_2H \rightarrow R_1H + R_2^{\cdot} \tag{4.7}$$

or

$$R_1^{\cdot} + M \rightarrow R_1 - M^{\cdot} \tag{4.8}$$

In principle, free radicals can also be detected by means of optical absorption measurements. This rather convenient method, which has been applied during flash photolysis investigations [32–36], is somewhat limited, since in most cases, it does not readily allow the identification of the radicals. The optical absorption method has proved quite powerful, however, in kinetic investigations of very fast reactions. Only a few of the many relevant cases, where macroradical reactions were studied via optical absorption, shall be mentioned here:

The spectrum of an aryloxyl radical of the structure

was recorded and its role in the Fries rearrangement was studied in 1,2 dichloroethane solution [32]. The transient biradical, occurring in the Norrish type II scission reaction, was detected when poly(phenylvinyl ketone) was irradiated in benzene solution [33, 36]:

Table 4.2   Free radicals generated during the photolysis of polymers (adopted from [30])

Polymer	Radical	Wavelength of Incident Light (nm)	Temperature during Irradiation (°C)
Polycarbonate	[benzene ring]—O• [benzene ring]—•	UV*)	−196
Polyethylene-terephthalate	$-CH_2-CH_2-O\bullet$	313	25
Polyvinyl-pyrrolidone	CH_2-CH_2 CH_2 $C=O$ N $-CH_2-\underset{\bullet}{C}-CH_2-$	UV	−196
Polyamide-6	$-CH_2-\underset{\bullet}{CH}-CH_2-$ $-CH_2-\underset{\bullet}{C}=O$	250−600	−196
Poly(2,6-di-methylphenylene oxide)	CH_3 [benzene ring]—O• CH_3	300	room temperature in benzene
Polystyrene	$-CH_2-\underset{\bullet}{C}-CH_2-$ [benzene ring]	250	−185
Polymethyl-methacrylate	$-CH_2-\underset{\bullet}{C}-CH_3$ $O=C-OCH_3$	253.7	25
	$\bullet CH_3$, $\bullet CHO$, $\bullet C-O-CH_3$ $\parallel$ O	253.7	−196

*) high pressure mercury lamp

$$-CH-CH_2-CH-CH_2 \rightarrow -CH-CH_2-\overset{\cdot}{C}-CH_2-$$

$$\begin{array}{cccc}
| & | & | & | \\
C=O^* & C=O & \cdot C-OH & C=O
\end{array}$$

(4.9)

$$-CH_2 + CH_2=C-CH_2-$$

$$\begin{array}{cc}
| & | \\
C=O & C=O
\end{array}$$

Moreover, optical absorption studies carried out recently on radicals of polystyrene [28], polyvinylbenzophenone [29] and homo- and copolymers of phenylisopropenylketone [33 (c)] might be mentioned.

Little will be said here about ionic species, since in most cases of interest photoionization is not expected to occur, i.e. at $\lambda_{inc} > 250$ nm. Nevertheless, evidence for the generation of ions during UV irradiation has been inferred from electrical conductivity measurements. Reportedly [37], photocurrents developed in polyethylene upon irradiating the polymer in the wave length range 180 to 360 nm. An unequivocal identification of the nature of the charge carriers was not possible, however. Electrons were assumed to act as charge carriers. *Binks* et al. [38] who stressed the importance of photoemission of electrons from the electrode, detected photoconductivity with various polymers such as polyethylene, polypropylene, polystyrene, cellulose acetate and poly(ethylene terephthalate). Photoionization according to reaction (4.5) has been found [39] for various systems with strong donor-acceptor interaction. The polymers consisted of heterocyclic or aromatic repeating units, e.g. poly-N-vinylcarbazole, polyvinylnaphthalene, polyvinylquinoline. Tetracyanoethylene, 1,3,5-trinitrobenzene, anthraquinone, 9,10-dichloroanthracene and similar compounds served as electron acceptors. Respective laser flash photolysis studies have revealed, recently, that the quantum yields for ion formation decrease as the molecular weight increases [40].

4.2.2   Energy Transfer and Energy Migration

Fig. 4.8 depicts possible pathways of a photon absorbed by a homopolymer, i.e. by a macromolecule composed of identical repeating units. Since each repeating unit contains the same chromophore, excitation energy can travel down the chain, provided orderly domains exist permitting the necessary interaction of neighboring chromophores or lifetimes of excited states are long enough to permit pendant groups to attain favorable geometrical positions. The latter is feasible in polymer solutions and in solid polymers at temperatures above the glass temperature range.

It is appropriate to distinguish *energy migration*, i.e. transfer of electronic excitation energy between like molecules, from *energy transfer* between unlike molecules [24]. It is noteworthy, that energy migration down the chain is a polymer-specific process.

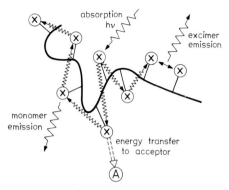

Fig. 4.8 Schematic illustration of intramolecular energy migration and energy transfer in polymers. Monomer and excimer emission are also shown

The reader might ask: How are energy transfer and energy migration related to polymer degradation? Actually, energy transfer is quite important, because, it can have a strong influence on polymer stability.

Principally, there are two modes of action with respect to energy transfer and polymer degradation

(I) *Sensitization:* $\qquad S + h\nu \to S^*$ (4.10a)

$\qquad\qquad\qquad\qquad S^* + P \to S + P^*$ (4.10b)

Excited polymer molecules P^* are generated via reaction (4.10b) and can undergo chemical reactions.

(II) *Protection:* $\qquad P + h\nu \to P^*$ (4.11a)

$\qquad\qquad\qquad\qquad P^* + Q \to P + Q^*$ (4.11b)

Excited polymer molecules P^* are "quenched" by an additive Q according to reaction (4.11b), which implies the inhibition of chemical reactions of P^*.

Before discussing characteristic cases, the mechanism of electronic energy transfer and conditions favoring it, will be discussed. These topics have recently been treated in a very comprehensible way by *Turro* [41].

Energy transfer processes, described generally by reaction (4.12)

$$D^* + A \to D + A^* \qquad (4.12)$$

(D = donor, A = acceptor, asterisk = electronically excited state)

are principally determined by reaction energetics. Although endothermic processes are possible, interesting cases pertain to exothermic reactions. In the former case the required activation energy is equal or greater than the endothermicity. Energy transfer occurs isoenergetically, i.e. the transition energies $D^* \to D$ and $A \to A^*$ must match perfectly. Fig. 4.9 allows the visualization of an exothermic transfer of electronic energy. Corresponding energy levels must exist in donor and acceptor molecules. Energetically allowed transitions are given by the spectral distribution of the donor emission, f_D, and the respective distribution of acceptor absorption, f_A. The spectral overlap integral

$$J = \int_{v}^{v'} f_D f_A \, dv \qquad\qquad (4.13)$$

provides a measure of the overlap of donor emission and acceptor absorption spectra (as indicated by the shaded area in Fig. 4.9). If $J = 0$, energy transfer is impossible. For $J \neq 0$, the magnitude of the rate constant of reaction (4.12) depends on the specific mechanism. Principally long and short range mechanisms can be distinguished.

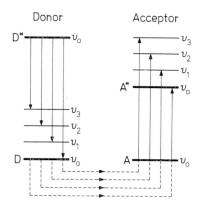

Fig. 4.9 Schematic illustration of corresponding transitions in exothermic energy transfer.
Upper part: energy levels in donor and acceptor becoming involved in the transfer. Lower part: emission spectrum of donor and absorption spectrum of acceptor. The spectral overlap, indicated by the shaded area, determines the magnitude of J in Eq. (4.13). For further information refer to [41]

Long range energy transfer processes can consist of emission-reabsorption reactions or so-called Coulomb interactions. In the former case acceptor molecules absorb light that is emitted by donor molecules. Coulomb interactions consist of dipole-dipole interactions. The rate constant for energy transfer via Coulomb interaction, k_{ET}, is given by the expression

$$k_{ET} \, (\text{Coul.}) \propto \frac{k_D^0 \, \varepsilon_A}{R_{DA}^6} \, J \qquad\qquad (4.14)$$

k_D^0 = rate constant for the emission $D^* \to D$
ε_A = absorption coefficient ($A \to A^*$)
R_{DA} = distance between the centers of the dipoles (A and D)
J = the spectral overlap integral

Short range interactions can be visualized as electron exchange interactions occurring if D^* and A approach each other closely enough. In other words, D^* and A are thought to collide so that their electron clouds overlap extensively, thus enabling electron exchange interaction to occur. The following expression holds for the rate constant of energy transfer in this case:

$$k_{ET} \text{ (electron exchange)} \propto J e^{-R_{DA}} \tag{4.15}$$

Table 4.3 shows several examples for energy transfer from a polymer to a low molecular weight acceptor. In principle, the transfer can occur in two modes from an electronically excited donor in its singlet state to an electronically excited acceptor in its singlet state or, analogously, from a triplet excited donor to a triplet excited acceptor. Triplet-triplet energy transfer is forbidden by Coulomb interactions but spin-allowed by the exchange mechanism. Experimentally both singlet-singlet and triplet-triplet transfer has been measured by quenching donor luminescence and/or sensitizing acceptor emission.

Table 4.3   Energy transfer in polymers

Donor	Acceptor	Mode of Transfer	R_{DA}^0 *) (Å)	Ref.
polystyrene	tetraphenyl-butadiene	singlet-singlet	40	42
polystyrene	cumene hydro-peroxide	singlet-singlet	11	43
polyvinylcarbazole	benzophenone	singlet-singlet	26	43
polyvinylnaphthalene	benzophenone	singlet-singlet	15	44
polyvinylnaphthalene	1,3-pentadiene	triplet-triplet	~15	46
polyphenyl-vinylketone	naphthalene	triplet-triplet	26	44
polyvinyl-methylketone	benzophenone	singlet-singlet	8	45
polyvinyl-methylketone	naphthalene	triplet-triplet	11	45
poly(styrene-co-vinylbenzophenone)	naphthalene	triplet-triplet	300	47

*) R_{DA}^0: critical distance between D^* and A, at which the probabilities for spontaneous deactivation and for energy transfer are equal

In Table 4.3, values of R_{DA}^0, the critical distance between D^* and A, at which the probabilities for spontaneous deactivation and for energy transfer are equal, are also given.

In several systems, such as polyvinylcarbazole/benzophenone, poly(phenylvinyl ketone)/naphthalene and poly(styrene-co-vinylbenzophenone)/naphthalene, experimentally determined R_{DA}^0 values are significantly larger than the calculated ones. It was therefore concluded that substantial energy migration occurs [43, 44, 47]. It is interesting to note

Table 4.4   Photolytic yields of homopolymers irradiated at room temperature in vacuo

Polymer	$\Phi(S)/\Phi(X)$	$\Phi(S)\times10^2$ [b]	$\Phi(X)\times10^2$ [c]	λ [a] (nm)	Volatile Products ($10^2\,\Phi$ in brackets)	Ref.
Polymethylmethacrylate	∞	1.2–3.9		253.7	CH_3OH (48); $HCOOCH_3$ (14) CO; H_2; CH_4; CO_2	48–52
Polyvinylacetate	1.4	6.6	4.7	253.7	CH_3COOH (1); CO_2 (0.65) CO (0.69); CH_4 (0.38)	53
Polyethyleneterephthalate	2.7	0.16	0.06	313	CO (0.06); CO_2 (0.02); $RCOOH$ (0.017)	54–55
Poly(methylvinyl ketone)		4.0		253.7	CH_4; CO; CH_3CHO	56
Poly(phenylvinyl ketone)	∞	6		313		57

[a] wavelength of incident light
[b] quantum yield for main-chain scission
[c] quantum yield for crosslinking

that excitation energy is quite efficiently transferred from polystyrene to cumene hydroperoxide. This finding suggests that the photooxidation of polystyrene can be strongly influenced by a sensitized photo-decomposition of incorporated hydroperoxide groups into reactive free radicals [43].

Additional information is available from [1, 6, 15, 41, 43].

4.3   Degradation in the Absence of Oxygen

In Table 4.4, quantum yields of photolysis products are listed for several polymers that were irradiated in the absence of oxygen at ambient temperature. It can be seen that the quantum yields for main-chain scission are lower than 10% ($\Phi(S) < 0.1$). In the case of polymethylmethacrylate, side chains are affected extensively as indicated by the high quantum yield for methanol ($\Phi = 0.48$). Quite frequently main-chain cleavage and intermolecular crosslinking occur simultaneously. In various cases $\Phi(S)/\Phi(X) < 4$, i.e. crosslinking dominates and gel is formed eventually (see also Section 5.2.2).

In Table 4.5, polymers are classified according to the predominant reaction.

Table 4.5   Predominant effects in the photolysis of solid polymers in the absence of oxygen

Main-Chain Scission (decrease of MW)	Crosslinking
Poly-α-methylstyrene	Polystyrene
Polymethylmethacrylate	Polymethylacrylate
Polyphenylvinylketone	Polyethylacrylate
	Polyvinylacetate
	Polyacrylonitrile
	Polyvinylbenzophenone
	Polyethyleneterephthalate
	Poly-2,2-propane-bis(4-phenylcarbonate)

The mechanism of the photolysis of most polymers has not yet been satisfatorily elucidated. To a large extent, this is because polymers usually contain impurities that absorb light more efficiently than the polymer itself and/or that act as energy acceptors or donors, thus giving rise to a photochemistry quite different from that in the "pure" polymer. Carbonyl groups play a prominent role among these impurities. Therefore, the photochemistry of ketone polymers was selected to demonstrate how light-induced chemical reactions proceed. As can be seen from Scheme 4.3, light absorbed by carbonyl groups can induce bond scission by either Norrish type I or type II processes. In the latter process a six-membered cyclic transition state is assumed, which decays by abstraction of a hydrogen atom from the C-atom in the γ-position. The resulting biradical cleaves rapidly according to reaction (b) in Scheme 4.3. Whereas type I processes generate lateral macroradicals, according to reaction (a), they lead to main-chain ruptures, if the carbonyl group is incorporated in the polymer backbone, as shown in reaction (c). In this case, type II processes also cause main-chain scissions. It is interesting to note that, in various cases, the fraction of absorbed photons, utilized for chemical transformation, depends significantly on polymer mobility. This is seen from Table 4.6. The results presented there suggest that the physical state strongly influences the magnitude of $\Phi(S)$. $\Phi(S)$ is rather low at room temperature for the copolymers poly(styrene-co-methylvinyl ketone) and poly(methylmethacrylate-co-methylvinyl ketone). Above the glass transition temperature, T_g, however, $\Phi(S)$ is much larger. A similar behavior has been found, when the polymers were irradiated in solution. The importance of molecular mobility derives from the fact that in type II processes the close approach of excited carbonyl groups to H atoms at γ-carbons is a prerequisite for reaction. Moreover, cage recombinations of radical pairs (produced, e.g., in type I processes) become less probable,

Scheme 4.3   Photochemical reactions in polymers containing ketone groups

Carbonyl in pendant groups:

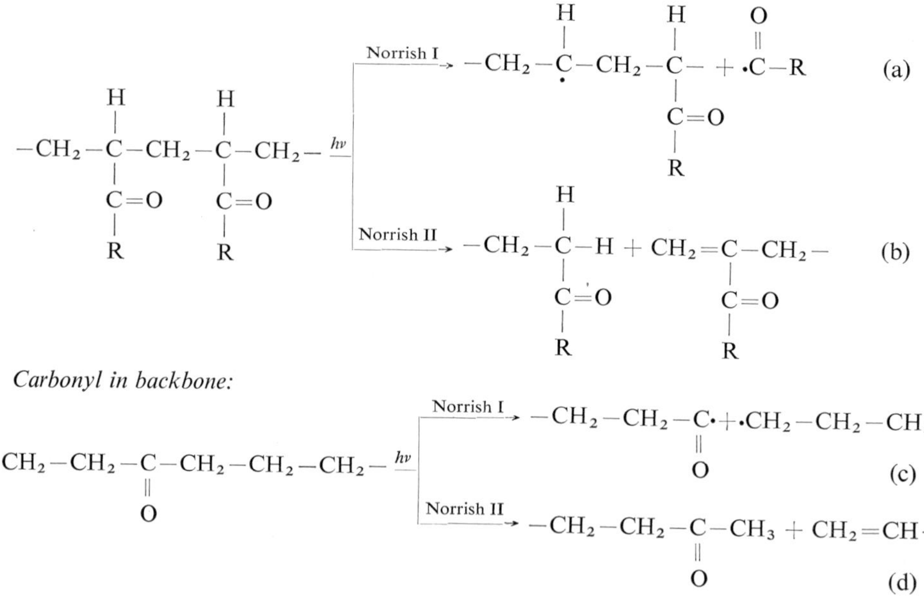

Carbonyl in backbone:

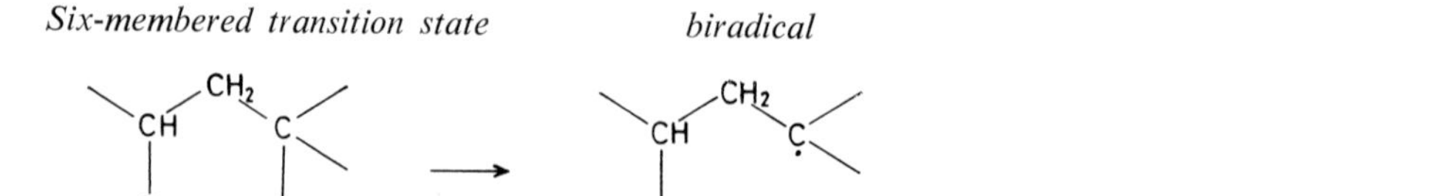

Six-membered transition state   *biradical*

'ance with this concept are the results reported
ᴄ. In this case, T_g lies at a temperature lower
ᴄ a temperature increase to values higher
ᴄnan the melting point of crystallites, should

ᴄm yields of main-chain rupture determined by irradiation
ᴄners at $\lambda = 313$ nm in vacuo

ᴄymer *)	$\Phi(S)$ $\times 10^2$	T_{irr} **) (°C)	T_g ***) (°C)	Ref.
Ethylene-CO	2.5	25	-110	58
	3.5	90	to -80	
St-PVK (9%)	4.4	27	ca. 90	57
	30	119−138		
MMA-MVK (7%)	1.9	27	105	57
	21	110−121		

  *) St: styrene; MMA: methylmethacrylate; PVK: phenylvinyl ketone;
     MVK: methylvinyl ketone
 **) temperature at irradiation
***) glass transition temperature

4.4 Photooxidation

4.4.1 Autoxidative Processes

As has been pointed out in the preceding section, in the absence of oxygen the quantum yields for chemical conversion are, in most cases, rather low. The situation can be quite different, however, if oxygen is present. Usually, free radicals are generated as transient species in photolytic processes. Since oxygen reacts readily with most free radicals, peroxyl radicals will be formed instead of products generated otherwise in the absence of oxygen. Photolysis, therefore, can give rise to autoxidative free radical chain reactions, the general principles of which have been dealt with in Section 1.3.2.

Photolysis, a special mode of initiation of autoxidation, is equivalent to other modes of initiation with regard to the end products formed. As far as the latter are concerned, it does not make any difference whether the initiation occurs via thermolysis, mechanical stress, chemical attack or via photolysis. Specific features of the various modes of initiation are discussed in the respective chapters of this book. A quite interesting phenomenon in photolytic oxidation of polymers is that additional chromophores are created during chain propagation. These chromophores can give rise to the initiation of new chain reactions upon prolonged irradiation and thus to rapid deterioration of the polymer.

According to the conventional mechanism of autoxidation

$$R_1 - R_2 \xrightarrow{hv} R_1^\bullet + R_2^\bullet \tag{4.16}$$

$$R^\bullet + O_2 \rightarrow RO_2^\bullet \tag{4.17}$$

$$RO_2^\bullet + RH \rightarrow ROOH + R^\bullet \tag{4.18}$$

hydroperoxide groups are formed in the propagation reaction. At wavelengths below 300 nm hydroperoxides are photolytically decomposed:

$$ROOH + hv \rightarrow RO\bullet + \cdot OH \tag{4.19}$$

Reaction (4.19) is considered to be very important in the photoinitiated oxidation of many commercial polymers. The latter contain peroxide groups as chemically bound impurities, originating from processing at elevated temperature in the presence of oxygen.

It has been suggested by *Geuskens* et al. [94] that, in the photolysis of polystyrene containing hydroperoxide impurities, main-chain scission could occur according to reaction (4.20):

$$\tag{4.20}$$

which is assumed to predominate, in rigid matrices, over reaction (4.19)

Apart from the monomolecular decomposition, according to reaction (4.19), a bimolecular mechanism can become operative at high local concentrations of hydroperoxide groups, as has been emphasized occasionally [62, 92, 93]:

$$P-O-O \cdots H-O-O-P \xrightarrow{hv} PO\bullet + H_2O + PO_2^\bullet \tag{4.21}$$

High local POOH concentrations (accumulation of POOH groups in clusters) can prevail, if the polymer under consideration had been subjected to autoxidation prior to photolysis (see Chapter 1). Alternatively, high local POOH concentrations can be attained at advanced stages of photo-initiated autoxidations.

Frequently, carbonyl groups are major constituents of the end products, being formed, for instance, in the termination reaction

$$\tag{4.22}$$

Carbonyl groups can also act as chromophores. Light absorption then gives rise to radical formation according to reaction (4.3), or alternatively, Norrish type I or type II reactions take place.

Because of the great technical importance of photooxidation of polymers, a great body of review articles and relevant special publications is available [3, 4, 7—9, 12, 60, 61]. Additional material is being continuously published.

4.4.2   Sensitized and Additive-Initiated Degradation

Recently the term sensitization has been used by photochemists to solely denote processes involving energy transfer which is subsequently followed by a chemical reaction or a physical process, such as luminescence *)

$$S^* + P \to S + P^* \tag{4.23}$$

$$P^* \overset{\text{chemical reaction}}{\underset{\text{physical deactivation}}{\Big|}} \begin{array}{l} \to P_1^\bullet + P_2^\bullet \quad (4.24\,\text{a}) \\ \to P + h\nu \quad (4.24\,\text{b}) \end{array}$$

In conjunction with sensitization, it is appropriate to discuss also processes such as hydrogen abstractions, that are initiated by electronically excited additives according to

$$S^* + PH \to HS\bullet + P\bullet \tag{4.25}$$

Sometimes excited initiator molecules react primarily with a low molecular weight compound:

$$S^* + RH \to HS\bullet + R\bullet \tag{4.26}$$

The free radicals produced in reaction (4.26) might subsequently attack the polymer

$$R\bullet + PH \to RH + P\bullet \tag{4.27}$$

For reasons of simplicity only hydrogen abstraction reactions were taken into consideration here, but other reactions, such as radical additions or ionic processes, could also occur.

Noteworthy are, moreover, processes originated by excited initiator molecules dissociating readily into free radicals after excitation:

$$S^* \to S_1^\bullet + S_2^\bullet \tag{4.28}$$

The radicals thus produced might attack the polymer.

A few examples are referred to in the following to illustrate sensitization and initiation.

Ketones and quinones, such as benzophenone, diacetyl, p-quinone, 1,4-naphthoquinone, 1,2-benzanthraquinone and 2-methylanthraquinone, have been frequently used in order to initiate or accelerate photodegradation. These compounds are capable of effectively absorbing UV light at $\lambda > 300$ nm and reacting usually, according to reaction (4.25). The excited states of benzoin and certain benzoin derivatives have, on the other hand, very short lifetimes (10^{-9} to 10^{-10} s) and decompose with relatively high quantum yields according to reaction (4.28).

It should be pointed out that, in the presence of oxygen, macroradicals produced via an indirect effect will be readily converted to peroxyl radicals, i.e. autoxidation is initiated as outlined in the previous section.

---

*) Notations used in reactions (4.23) to (4.28):
  S = additive; P = macromolecule; PH and RH = macromolecule and low molecular weight molecule, respectively, capable of acting as hydrogen donor; $P\bullet$ and $R\bullet$ = free radicals; asterisk = electronically excited state

In Scheme 4.4, the mechanism of the benzophenone initiated photodegradation of poly-acetaldehyde is shown [73].

Scheme 4.4   Benzophenone initiated photodegradation of polyacetaldehyde

(a)

(b)

(c)

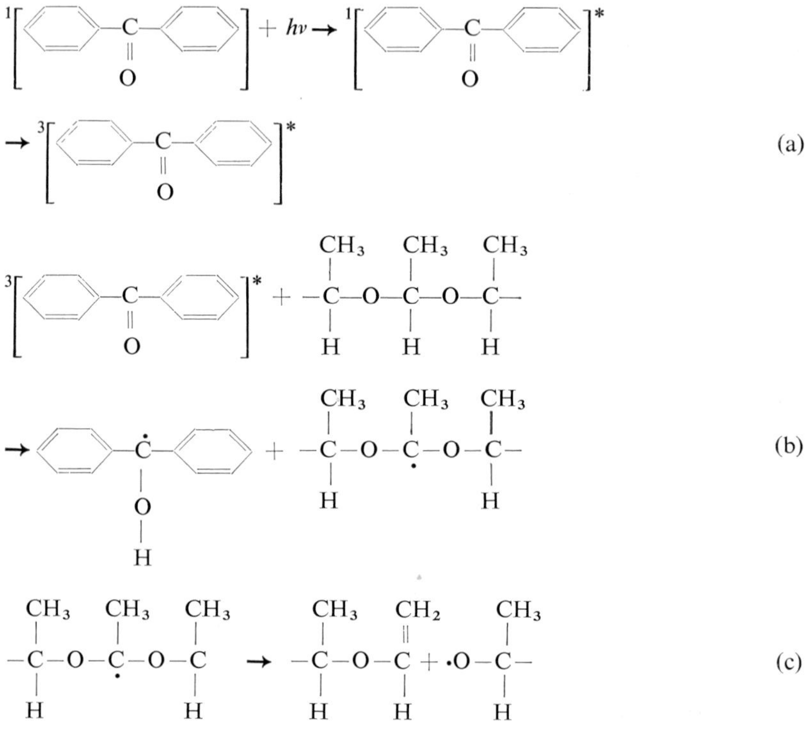

A significant acceleration of the rate of main-chain scission in polymethylmethacrylate and bisphenol-A polycarbonate is caused by anhydrous $FeCl_3$, according to *Mikheyev* et al. [63]. It is assumed that, after the light has been absorbed by $Fe^{3+}Cl^-$ ion pairs distributed in the polymer matrix, chlorine atoms subsequently formed attack the polymer:

$$Fe^{3+}Cl^- + h\nu \rightarrow [Fe^{2+}Cl] \rightarrow Fe^{2+} + Cl\cdot \qquad (4.29)$$

Of practical interest is the fact that titanium dioxide pigments (anatase *) can remarkably enhance the photodegradation of polyamides [64—66]. Manganese(II) compounds were found to counteract photodegradation. Thus, it is advisable to precoat titania with a manganese compound before application [67—68].

It has been suggested that sensitization mechanisms involving reaction (4.23) are responsible for the accelerator action of polycyclic aromatic hydrocarbons (PA), (e.g., naphthalene(I), phenanthrene(II),and hexahydropyrene(III), in polyethylene and polypropylene).

---

*) rutile is relatively inactive

Biphotonic processes, which give rise to excited carbonyl groups incorporated in the polymer, are thought to occur in these cases [69−72]:

$$^1PA \xrightarrow{hv} {}^1PA^* \rightarrow {}^3PA^* \qquad (4.30)$$

$$^3PA^* + hv'_* \rightarrow {}^3PA^{**} \qquad (4.31)$$

$$^3PA^{**} + {}^1\left[\begin{array}{c} \diagup \\ \diagdown \end{array}C{=}O\right] \rightarrow {}^1PA + {}^3\left[\begin{array}{c} \diagup \\ \diagdown \end{array}C{=}O\right]^* \qquad (4.32)$$

The subsequent decomposition reactions have been described in Scheme 4.3.

Noteworthy is, furthermore, the sensitized decomposition of hydroperoxide groups by triplet excited ketones:

$$^3\left[\begin{array}{c} \diagup \\ \diagdown \end{array}C{=}O\right]^* + POOH \rightarrow {}^1\left[\begin{array}{c} \diagup \\ \diagdown \end{array}C{=}O\right] + PO^{\bullet} + {}^{\bullet}OH \qquad (4.33)$$

It has been suggested that this reaction becomes important at an advanced stage of photooxidation, i.e. after accumulation of hydroperoxide groups to a sufficiently high concentration [62].

For additional literature references concerning accelerated degradation of polymers the reader may refer to [74, 75].

4.4.3 Stabilization

As no polymer is capable of withstanding prolonged exposure to solar radiation, stabilization of commercial polymers is extremely important. It was indicated in the preceding sections that a great deal of light-induced damage to polymers is due to free radical oxidative chain reactions. Antioxidants of the chain terminator type should, therefore, act also as photostabilizers, since − apart from initiation reactions − similar mechanism are operative both in photo-oxidation and in thermal oxidation (see Section 2.6).

The characteristics of photodegradation, however, permit the application of additional methods involving both photon absorption and transfer of electronic energy. Generally, photostabilizers for polymers are classified according to their mode of action: (a) light screeners, (b) UV light absorbers, (c) excited state quenchers and (d) antioxidants. A variety of stabilizers are multifunctional, mostly bifunctional.

Screening is the most obvious method for protecting articles from light absorption. Painting, which usually serves as a means of protection, is not applicable to most plastic materials owing to the incompatibility of polymers with dyes. Incorporation of pigments works quite well in many cases, because most pigments absorb not only visible but also harmful UV light. Furthermore, light absorption in pigmented systems is restricted to a thin surface layer and the underlying regions remain unaffected. Carbon black is one of the most effective pigments. The applicability of pigments is limited, unfortunately, because of compatibility problems, that exist in many cases. Moreover, as pointed out in Section 4.4.2, certain pigments accelerate photodegradation.

Typical stabilizers acting as screeners, UV absorbers, and/or quenchers or antioxidants, are listed in Table 4.7.

Both screeners and UV absorbers operate by the same mechanism: they absorb UV light with a high extinction coefficient. The absorbed energy is harmlessly dissipated, i.e. to an overwhelming degree converted to heat. The polymer is not or only slightly damaged, since only a small portion of incident light can reach its chromophoric sites. 2-hydroxybenzophenones, phenylsalicylates, resorcinol esters, 2-hydroxy-benzotriazoles and coumarine derivatives are the most widely used UV absorbers.

Table 4.7   Photostabilizers (selected examples).
Compilations of stabilizers can be found in [14, 76]

Compound class	Compound		Principal Mode of Action
Pigments	Carbon Black, ZnO, MgO, CaCO$_3$, BaSO$_4$, Fe$_2$O$_3$		UV Screener
2-Hydroxy-benzo-phenones		R_1 = H, alkyl R_2 = H, alkyl, phenyl R_3 = H, butyl R_4 = H, butyl	UV Absorber
		R = alkyl	
Phenylsali-cylates			UV Absorber
Benzo-triazoles		R = H, alkyl	UV Absorber

Table 4.7 (Continuation)

Compound class	Compound		Principal Mode of Action
Nickel chelates	$\left(\begin{array}{c} R \\ \diagdown \\ N-C-S- \\ \diagup \\ R \end{array} \right)_2 Ni$ (with S double bonded to C)		Quencher
	$\left(R-O-\overset{\overset{\displaystyle S}{\|}}{\underset{\underset{\displaystyle R}{\|}{O}}{P}}-S- \right)_2 Ni$	$R = $ alkyl	
	Ni chelate with CH_3, $C=N$, R, O aromatic rings and $N=C$, CH_3, R groups		

- -

| Hindered Amine | Bis-piperidine structure: $OCO(CH_2)_8OCO$ bridging two 2,2,6,6-tetramethylpiperidine rings (H, CH_3, CH_3, N–R) | $R = H$, alkyl | Antioxidant |

It is interesting to note that 2-hydroxy-benzophenones act as stabilizers whereas benzo-phenone and its derivatives accelerate photodegradation. The different behavior can be explained on the basis of intramolecular hydrogen bonding in the excited state. A rapid deactivation can be visualized according to Scheme 4.5, if I is directly excited and the equilibrium between the tautomers II and III is strongly shifted to the side of III, which very efficiently undergoes radiationless deactivation. Stabilizer action involving triplet energy transfer is feasible if the lifetime (with respect to radiationless deactivation) of IV or V is short. Alternatively, if V is the dominating form in the triplet equilibrium IV ⇌ V, its unreactive nature can be due to the fact that the lowest lying state is of π, π^* character. The lowest lying triplet state of benzophenone is of n, π^* character and is chemically quite reactive.

The protecting properties of salicylates and 2-hydroxy-benzotriazoles are likely to be understood on the basis of similar mechanisms [107–110].

Scheme 4.5   Modes of stabilizer action of 2-hydroxy-benzophenone

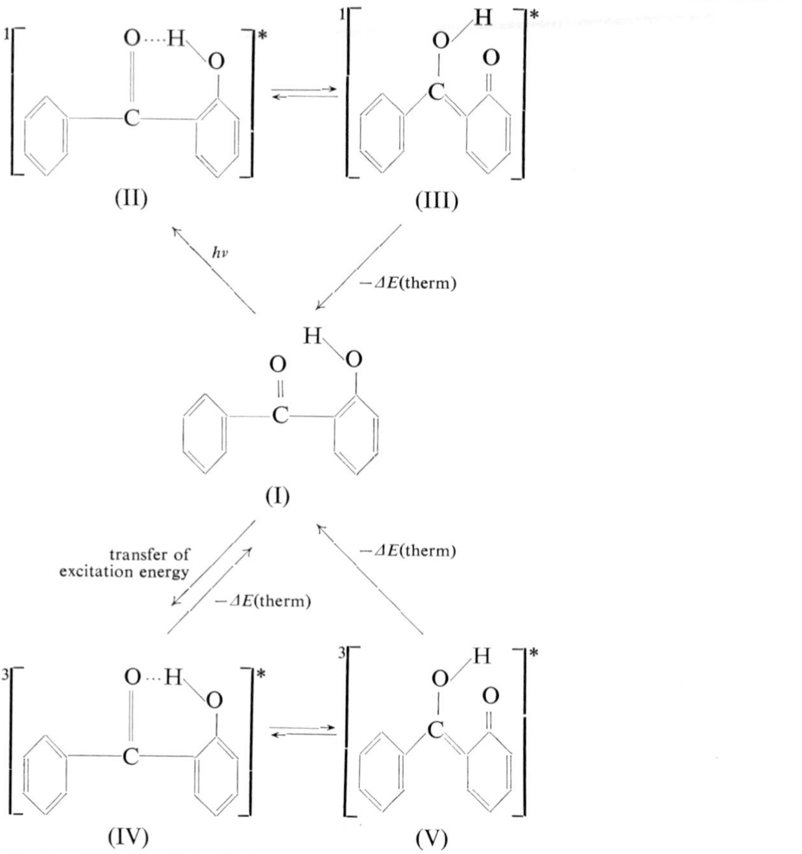

In the case of quenchers, the mode of action is a passive one. The light is absorbed by the chromophores of the polymer. Chemical deactivation via bond rupture and energy transfer to the quencher are competing processes:

$$P^* \quad \longrightarrow \quad R_1 \cdot \; + \; R_2 \cdot \tag{4.34a}$$

$$ \quad \xrightarrow{\;Q\;} \quad P \; + \; Q^* \tag{4.34b}$$

Obviously, the harmless deactivation of Q^* is a prerequisite for the applicability of a quencher as a stabilizer.

The importance of quenchers added to plastics derives mainly from their ability to interact with excited carbonyl chromophores. Carbonyl groups are almost ubiquitous in thermoplastics, especially in polyolefins.

Quenchers currently in use are complexes or chelates of transition metals (three nickel chelates are shown in Table 4.7). By laser flash photolysis, it has been demonstrated that a number of Ni(II) chelates effectively quench triplet-excited benzophenone [77].

The quenching effectivity depends on the magnitude of the corresponding donor and acceptor energy levels (see Section 4.2.2). Table 4.8 presents singlet and triplet energies of carbonyl chromophores incorporated in polystyrene and polypropylene, and also those of a nickel chelate stabilizer. The latter was found to act in polystyrene merely by screening, in polypropylene, however, by both screening and quenching [78, 79]. This becomes intelligible if energy transfer proceeds via triplet states, because $T_1(CO/PSt)$ $< T_1(\text{Ni-chelate}) < T_1(CO/PP)$. Exothermic energy transfer to the Ni chelate is possible only in the polypropylene matrix.

Table 4.8 Singlet and triplet energies of carbonyl chromophores and of a nickel chelate (2, 2'-thio-bis(4-t-octylphenolato)-n-butylamine nickel) [78]

Compound	Singlet S_1 (eV)	Triplet T_1 (eV)
$\diagup\!\!\!{>}C{=}O$ in Polystyrene	3.60	3.14
$\diagup\!\!\!{>}C{=}O$ in Polypropylene	4.09	3.66
Ni chelate	3.60	3.35

Typical light stabilizers of the antioxidant type are hindered amines shown at the end of Table 4.7 [75, 80, 81]. They significantly inhibit the photooxidation of polypropylene as shown in Fig. 4.10 (refer to Chapter 2.6 for the mechanism). The oxidative process is assumed to be initiated via a biphotonic mechanism involving α, β -unsaturated carbonyl groups [80]. The absorption of the first photon causes a shift of the double bond:

$$-\overset{|}{\underset{|}{C}}-\overset{|}{\underset{|}{C}}{=}C-\overset{\|}{C}- \xrightarrow{h\nu} -\overset{|}{\underset{|}{C}}{=}C-\overset{|}{\underset{|}{C}}-\overset{\|}{C}- \qquad (4.35)$$

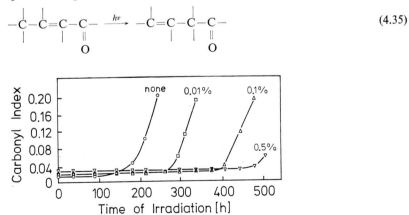

Fig. 4.10 Photooxidation of commercial polypropylene in the absence and presence of a hindered amine; for formula see: Table 4.7. (According to *Allen* and *McKellar* [80])

Subsequently, bond scission occurs via type I or type II processes upon the absorption of an additional photon.

It should be pointed out again, that many stabilizers are multifunctional in their mode of operation, and that the application of stabilizers still comprises a series of problems, such as compatibility of the stabilizer with the polymer, diffusion of the stabilizer in the polymeric matrix and photodecomposition of the stabilizer.

For additional information the reader is referred to [1, 2, 7, 9, 16, 75].

4.4.4   Singlet Oxygen Reactions

The ground state of molecular oxygen is a triplet state, with two unpaired electrons. Ground state oxygen readily reacts with free radicals, and electronically excited singlet and triplet states of molecules of different chemical nature. Table 4.9 presents the electronic state properties of molecular oxygen.

Table 4.9   Electronic state properties of molecular oxygen (O_2) (according to *Kearns* [91])

State	Electronic Configuration of Highest Occupied MO's		Relative Energy (kJ/mol)	Lifetime (in liquid systems) (s)
$^3\Sigma_g^-$ (ground state)	↑ (↓)	↑ (↓)	0	
$^1\Delta_g$ (1st excited state)	↓↑	—	94	ca. 10^{-4}
$^1\Sigma_g^+$ (2nd excited state)	↑	↓	157	ca. 10^{-10}

The first excited state, $^1O_2^*$, is a singlet state, which is of chemical importance owing to its rather long life time. $^1O_2^*$, is generated by energy transfer processes (quenching) if the difference in the corresponding energy levels exceeds 94 kJ/mol. Therefore, the generation of $^1O_2^*$ via reaction (4.36)

$$^3X^* + {}^3O_2 \rightarrow {}^1X + {}^1O_2^* \tag{4.36}$$

is feasible for many dyestuffs, carbonyl groups and polynuclear aromatics, e.g.:

$$\left[{}^3\!\!\!\diagdown\!\!C\!=\!O \right]^* + {}^3O_2 \rightarrow \left[{}^1\!\!\!\diagdown\!\!C\!=\!O \right] + {}^1O_2^* \tag{4.37}$$

Systematic studies of the reactions of low molecular weight model compounds with $^1O_2^*$, generated by microwave discharge or via the decomposition of the triphenyl-phosphite/ozone adduct, revealed the following:

$^1O_2^*$ is quite reactive towards unsaturated substances such as olefins and dienes [82].

The reaction results in the introduction of hydroperoxide groups, e.g.

$$\text{(4.38)}$$

On the other hand, $^1O_2^*$ was found to be unreactive towards saturated hydrocarbons [83].

Photochemical studies, taking advantage of the possibility to generate $^1O_2^*$ via reaction (4.36) (e.g. with the aid of α, β, γ, δ-tetraphenylporphyrin), showed that polymers with unsaturation in pendant groups or in the backbone exhibited reactivity. Polymers not containing olefinic unsaturations were unreactive: e.g. polyvinylchloride, polystyrene, and polyethylmethacrylate [84—86]. It is worth pointing out that, by laser flash photolysis studies, the rate constants of the reaction of $^1O_2^*$ with unsaturated hydrocarbons have been determined as $10^3 - 10^4$ l/mol s, which implies only a moderate reactivity [87].

It can be concluded that singlet oxygen reactions can contribute in certain cases to the initiation of photooxidative processes. From the present state of knowledge, it can be inferred that singlet oxygen plays a role only in polymers containing olefinic unsaturations.

The suggestion by Trozzolo and Winslow that singlet oxygen could participate in the photooxidation of polymers [88], generated some controversy, but this appears to calm down as results from systematic studies become available. The topic has been concisely reviewed [89]. For additional information refer to [7, 90].

4.5   Applications

4.5.1   Polymers with Predictable Lifetime

A large part of the litter problem in holiday resorts and recreation areas, such as beaches, consists of polymeric substances, mostly packaging material (empty bottles and containers or plastic bags etc.). These items, thrown away carelessly, frequently exhibit a significant resistance to attack by microorganisms and to irradiation with sunlight. There are additional cases, especially concerning various applications in agriculture and gardening (e.g. plastic mulches on vegetable fields), where it would be highly desirable to have available procedures allowing cheap polymer decomposition. Photodegradation by sunlight would be most convenient for this purpose.

In this section, emphasis will be given to photochemical methods for the intended decomposition of polymers, whereas methods based purely on microbial attack will be discussed in some detail in Chapter 6, devoted to biodegration.

Essentially, two methods have been developed to control and predict lifetimes of plastics:

a) Introduction of ketone groups into polymers [95]. The relatively high photodegradability of ketone polymers has been utilized for the production of a variety of plastics to be decomposed by the UV portion of solar radiation. In order to provide acceptable rates of degradation, it is necessary, in many cases, to incorporate only a small percentage of carbonyl groups into the polymer. In the case of polystyrene, e.g., less than 1% of carbonyl containing moieties is sufficient to guarantee decomposition into small frag-ments within a few weeks. The method works with several other polymers: polyethylene, polypropylene, polymethylmethacrylate, polyacrylonitrile, polymethacrylonitrile, and certain copolymers [95]. With vinyl polymers, ketone group containing comonomers can easily be incorporated by copolymerization. In the case of condensation polymers, such as polyamides and polyesters, ketone group containing difunctional monomers must be used for the preparation of the polymer. Several industrial companies in different countries have developed processes for the commercial production of ketone group containing plastics. Although the decomposition starts as soon as the plastic is exposed to sunlight, there is a certain delay period before a breakdown of important physical properties occurs. As the plastic material is macroscopically broken down and as its polymer chains split, on a molecular level, it becomes more and more susceptible to biological degradation, i.e. to microbial attack.

b) Delayed action of added photoactivators [96]. In Fig. 4.11 a plot of the oxygen uptake vs. the irradiation time shows how an added substance exhibits a pronounced anti-oxidant action for a certain period, after which a remarkable acceleration in the oxidation rate, indicated by an exponential increase, occurs. This behavior is characteristic of a catalytic destruction of hydroperoxide groups formed during processing in polymers. At the end of the induction period, the additive is used up and decomposed. Subsequently, the decomposition products catalyze the photodegradation process.

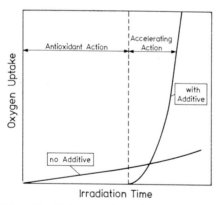

Fig. 4.11   Delayed action of photoactivator. Schematic illustration of the oxygen uptake as a function of time of UV irradiation in the absence and presence of an additive

Typical additives possessing the described properties are Fe(III)-acetylacetonate(I) and Fe(III)-2-hydroxy-4-methylacetophenone oxime(II) [97].

The oxidative degradation of polyethylene has been investigated thoroughly with respect to possible applications of this method, which is said to render the polymer biodegradable. Owing to the presence of iron ions the disintegration of the polymer is reported to continue in the dark, after irradiation, even if the material is buried in soil or composted [96].

4.5.2   Photoresists

Technologies for the production of solid state electronic devices, such as integrated circuits, strongly depend on appropriate organic polymers being resistant to attack by etching agents. In other words, resists are organic materials used in the manufacture of microelectronic devices to protect underlying substrates during etching. As is shown in Fig. 4.12, there are resists acting in a positive mode upon irradiation, resulting in an

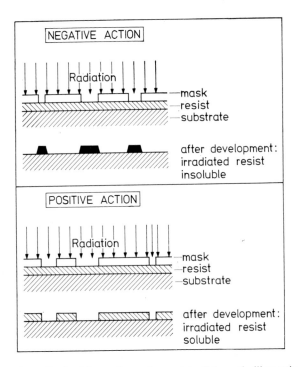

Fig. 4.12   Positive and negative resists. Schematic illustration of mode of action

increase in their solubility, and resists acting in a negative mode, resulting in insolubilization. Photoresists are used in combination with masks permitting exposure of the resist material at unprotected areas only. Positive resist action can be achieved by using polymers that undergo main-chain scission upon irradiation. Conversely, negative resist action is commonly based on intermolecular crosslinking.

Conventional positive photoresists operate on mechanisms based on the destruction of dissolution inhibitors. A typical base-insoluble composite system, consisting of a novolac resin and an o-quinone diazide derivative, becomes base-soluble upon irradiation, due to a photochemical rearrangement of the additive [98]. Modern requirements in fine line lithography aiming at minimum features of less than 1 μm, were an incentive for new developments [99] involving polymer systems undergoing main-chain scission upon irradiation with UV light at wave lengths between 200 and 300 nm. Polymethylmethacrylate with and without sensitizer and polymethylisopropenylketone [100] have been used, for example.

Poly(olefin sulfones), i.e. alternating copolymers of the general structure

$$\left[-R-\underset{\underset{O}{\|}}{\overset{\overset{O}{\|}}{S}}- \right]_n$$

exhibit interesting positive resist properties when used as electron resists (see Chapter 5). Because of the low-lying absorption band of the sulfone group, aliphatic polysulfones are not interesting from the photochemical point of view. A different situation is met, however, with aromatic polysulfones, such as poly(styrene sulfone) or poly(styrene-co-acenaphthylene sulfone) [103]. They have absorption peaks between 250 and 300 nm and possess high sensitivities.

A variety of negative photoresists, consisting of polymers containing reactive groups, are known. Frequently they are used in conjunction with photosensitizers. Typical examples of such polymers are cis-1,4-polyisoprene and cis-1,4-polybutadiene [102].

Additional negative resists were proposed recently [101], for example azide polymers such as polyvinylbenzylazide and poly(vinylbenzene sulfonyl azide) [104]. The subject has been reviewed recently by *Bowden* and *Thompson* [99].

4.5.3   Residue-free Decomposition with the Aid of Lasers

Actually, this application involves essentially thermal degradation, because continuous irradiation of a polymer sample with a high intensity laser beam causes a very rapid increase of temperature in the irradiated area resulting in thermal decomposition of the polymer into gaseous products (vaporization). Argon and krypton lasers are quite appropriate for this purpose, because a number of output wave lengths can be selected over the spectral region of 350 to 750 nm.

Laser decomposition of polymers is also applicable for the production of functional and decorative devices [105]. Another application pertains to the vaporization of polymer samples directly in the source cell of a mass spectrometer. Thus, additives, or traces of impurities down to the range of a few ppm can be readily detected [106].

References to Chapter 4

[1] *W. Schnabel* and *J. Kiwi*, "Photodegradation" in *H. H. G. Jellinek* (ed.), "Aspects of Degradation and Stabilization of Polymers", Elsevier (1978).

[2] *V. Ya. Shlyapintokh*, "Photokhimičeskie prevračeniya i stabilizatsiya polymerov", Izdatelstov Khimiya, Moscow (1979).

[3] *N. S. Allen* and *J. F. McKellar*, "The Role of Luminescent Species in the Photooxidation of Commercial Polymers" in *N. Grassie* (ed.), "Developments in Polymer Degradation-2", Appl. Science Publ., London (1979).

[4] *R. Arnaud* and *J. Lemaire*, "Photocatalytic Oxidation of Polyolefins" in *N. Grassie* (ed.), "Development in Polymer Degradation-2", Appl. Science Publ., London (1979).

[5] *G. Geuskens* (ed.), "Degradation and Stabilization of Polymers", Appl. Science Publ., London (1975).

[6] *G. Geuskens*, "Photodegradation of Polymers", in *C. H. Bamford* and *C. F. H. Tipper* (eds.), "Comprehensive Chemical Kinetics", Vol. 14, Elsevier, New York (1975).

[7] *B. Rånby* and *J. F. Rabek*, "Photodegradation, Photooxidation and Photostabilization", Wiley-Interscience, London (1975).

[8] *B. Sedláček* (ed.), "Mechanism of Inhibition Processes in Polymers: Oxidative and Photochemical Degradation", J. Polym. Sci. Polym. Symposium 40 (1973).

[9] *A. M. Trozzolo*, "Stabilization Against Oxidative Photodegradation" in *W. L. Hawkins* (ed.), "Polymer Stabilization", Wiley-Interscience, New York (1972).

[10] *K. Tsuji*, "ESR Studies of Photodegradation of Polymers", Adv. Polym. Sci. 12, 131 (1973).

[11] *N. Grassie*, "Degradation" in *A. D. Jenkins* (ed.) "Polymer Science", Vol. 2, North-Holland, Amsterdam (1972).

[12] *O. Cichetti*, "Mechanisms of Oxidative Photodegradation and UV Stabilization of Polyolefins", Adv. Polym. Sci. 7, 70 (1970).

[13] *R. F. Reinisch* (ed.) "Photochemistry of Macromolecules", Plenum Press, New York (1970).

[14] *D. Phillips*, "Polymer Photochemistry", Photochem. 1, 441 (1970); 3, 3, and 807 (1972); 4, 919 (1973); 5, 691 (1974); 6, 659 (1975); 7, 505 (1976).

[15] *R. B. Fox*, "Photophysical Processes and their Role in Polymer Photochemistry", Pure Appl. Chem. 30, 87 (1972).

[16] *J. Voigt*, "Die Stabilisierung der Kunststoffe gegen Licht und Wärme", Springer, Berlin (1966).

[17] (a) *F. T. Arecchi*, *F. O. Schultz-Dubois* (eds.), "Laser Handbook", North-Holland, Amsterdam (1976);
(b) *K. L. Kompa*, "Chemical Lasers", Fortschr. Chem. Forsch., Springer, Berlin (1973);
(c) *R. W. F. Gross* and *J. F. Bott* (eds.), "Handbook of Chemical Lasers", Wiley, New York (1976);
(d) *U. Köpf*, "Laser in der Chemie", Salle/Sauerländer, Frankfurt/M. (1979).

[18] *W. Koechner*, "Solid State Laser Engineering", Springer, Berlin (1977).

[19] *C. K. Rhodes* (ed.), "Excimer Lasers", Springer, Berlin (1979).

[20] *J. G. Calvert* and *J. N. Pitts, Jr.*, "Photochemistry", Wiley, New York (1966).

[21] *R. O. Kan*, "Organic Photochemistry", McGraw-Hill, New York (1966).

[22] *D. C. Neckers*, "Mechanistic Organic Photochemistry", Reinhold, New York (1967).

[23] *N. J. Turro*, "Molecular Photochemistry", Benjamin, New York (1967).

[24] *J. B. Birks*, "Photophysics of Aromatic Molecules", Wiley-Interscience, London (1970).

[25] *N. J. Turro*, "Modern Molecular Photochemistry", Benjamin/Cummings, Menlo Park, Cal. (1978).

[26] *H. G. O. Becker* et al. (Autorenkollektiv), "Einführung in die Photochemie", VEB Deutscher Verlag der Wissenschaften, Berlin (1976).

[27] (a) *K. P. Ghiggino*, *R. D. Wright* and *D. Phillips*, J. Polym. Sci., Phys. Ed. 16, 1499 (1978);

(b) *S. W. Beavan*, *J. S. Hargreaves* and *D. Phillips*, "Photoluminescence in Polymer Science", Adv. Photochem. 11, 207 (1978).

[28] *S. Tagawa* and *W. Schnabel*, Makromol. Chem., Rap. Comm. 1, 345 (1980),

[29] *W. Schnabel*, Makromol. Chem. 180, 1487 (1979).

[30] *K. Tsuji*, "ESR Study of Photodegradation of Polymers", Adv. Polym. Sci. 12, 131 (1973).

[31] *B. Rånby* and *J. F. Rabek*, "ESR Spectroscopy in Polymer Research", Springer, Berlin (1977).

[32] *J. S. Humphrey Jr.*, *A. R. Shultz* and *D. B. G. Jaquiss*, Macromolecules 6, 305 (1975).

[33] (a) *J. Kiwi* and *W. Schnabel*, Macromolecules 8, 430 (1975) and 9, 468 (1976);
     (b) *G. Beck*, *G. Dobrowolski*, *J. Kiwi* and *W. Schnabel*, Macromolecules 8, 9 (1975);
     (c) *I. Naito*, *R. Kuhlmann* and *W. Schnabel*, Polymer 20, 165 (1979).

[34] *N. S. Allen*, *D. Wilson* and *J. F. McKellar*, Makromol. Chem. 178, 1 (1978).

[35] *J. Faure*, *J. P. Fouassier*, *D. L. Lougnot* and *R. Salvin*, Eur. Polym. J. 13, 891 (1977); J. Photochem. 5, 13 (1976).

[36] *R. D. Small* and *J. C. Scaiano*, J. Phys. Chem. 81, 828, 2126 (1977); J. Photochem. 6, 453 (1976/77); Macromolecules 11, 840 (1978).

[37] (a) *T. Tanaka* and *Y. Inuishi*, Jap. J. Appl. Phys. 6, 1371 (1971);
     (b) *L. A. Vermeulen*, *H. A. Wintle* and *D. A. Nicodemo*, J. Polym. Sci. A-2, 9, 543 (1971);
     (c) *D. K. Dasgupta* and *M. E. Tindell*, J. Phys. Appl. Phys. D 5, 1368 (1972);
     (d) *T. Mitzutani*, *Y. Takai* and *M. Ieda*, Jap. J. Appl. Phys. 12, 1553 (1973);
     (e) *B. Andress*, Koll. Z. Z. Polym. 252, 650 (1974).

[38] *A. E. Binks*, *A. G. Campbell* and *A. Sharples*, J. Polym. Sci. A-2, 8, 529 (1970).

[39] *H. Hoegle*, J. Phys. Chem. 69, 755 (1965).

[40] *H. Masuhara*, *S. Ohwada*, *N. Mataga*, *A. Itaya*, *K. Okamoto* and *S. Kusabayashi*, Chem. Phys. Lett. 59, 188 (1978).

[41] *N. J. Turro*, "Energy Transfer Processes", Pure and Appl. Chem. 49, 405 (1977).

[42] *L. J. Basile*, Trans. Farad. Soc. 42, 3163 (1965).

[43] *G. Geuskens* and *C. David*, "New Aspects of Energy Transfer Phenomena in High Polymer Systems Including Degradation Phenomena", Pure and Appl. Chem. 49, 479 (1977).

[44] *C. David*, *W. Demarteau* and *G. Geuskens*, Europ. Polym. J. 6, 1397 and 1405 (1970).

[45] *C. David*, *N. Putman*, *M. Lempereur* and *G. Geuskens*, Europ. Polym. J. 8, 409 (1972).

[46] *C. David*, *M. Lempereur* and *G. Geuskens*, Europ. Polym. J. 8, 417 (1972).

[47] *C. David*, *V. Naegelen*, *W. Piret* and *G. Geuskens*, Europ. Polym. J. 11, 569 (1975).

[48] *A. R. Shultz*, Z. Phys. Chem. 65, 967 (1961).

[49] *A. Charlesby* and *D. K. Thomas*, Proc. Roy. Soc. London Ser. A 269, 104 (1962).

[50] *R. B. Fox*, *L. G. Isaacs* and *S. Stokes*, J. Polym. Sci. Part A 1, 1079 (1963).

[51] *D. G. Gardner* and *L. M. Epstein*, J. Chem. Phys. 35, 1653 (1961).

[52] *V. Ya. Shlyapintokh* and *V. I. Goldenberg*, Eur. Polym. J. 10, 679 (1974).

[53] *G. Geuskens*, *M. Borsu* and *C. David*, Eur. Polym. J. 6, 959 (1970); 8, 883, 1347 (1970).

[54] *M. Day* and *D. M. Wiles*, J. Appl. Polym. Sci. 16, 203 (1972).

[55] *F. B. Marcotte*, *D. Campbell*, *J. A. Cleaveland* and *D. T. Turner*, J. Polym. Sci. Part A-1, 5, 481 (1967).

[56] *J. E. Guillet* and *R. G. W. Norrish*, Proc. Roy. Soc. London, Ser. A, 233, 153 (1955).

[57] *E. Dan* and *J. E. Guillet*, Macromolecules 6, 230 (1973).

[58] *G. H. Hartley* and *J. E. Guillet*, Macromolecules 1, 165 and 413 (1968).

[59] *Y. Amerik* and *J. E. Guillet*, Macromolecules 4, 375 (1971).

[60] *N. A. Weir*, "The Effect of Photo-Degradation and Photo-Oxidative Degradation on the Dielectric Properties of Polystyrene", in "Developments in Polymer Degradation-1", *N. Grassie* (ed.), Appl. Science Publ. London (1977).

[61] *G. Scott*, "The Role of Peroxides in the Photo-Degradation of Polymers" in "Developments in Polymer Degradation-1", *N. Grassie* (ed.), Appl. Science Publ., London (1977).

[62] *G. Geuskens* and *C. David*, "The Photooxidation of Polymers. A Comparison with Low Molecular Weight Compounds", Pure and Appl. Chem. 51, 233 (1979).

[63] (a) *D. Ya. Toptygin, G. B. Pariiskij, Ye. Ya. Davidov, O. A. Ledneva* and *Yu. A. Mikheyev,* Vysokomol. Soedin. Ser. A 14, 1534 (1972);
(b) *Yu. A. Mikheyev, G. B. Pariiskij, V. F. Shubnyakov* and *D. Ya. Toptygin,* Kim. Vys. Energ. 5, 77 (1971).
[64] *H. A. Taylor, W. C. Tincher* and *W. F. Hammer,* J. Appl. Polym. Sci. 14, 171 (1970).
[65] *H. G. Voelz, G. Kaempf* and *H. G. Fitzky,* Fabe Lack 78, 1037 (1972).
[66] *G. S. Egerton,* Nature (London) 204, 1153 (1964).
[67] *N. S. Allen, J. F. McKellar, G. O. Phillips* and *C. B. Chapman,* J. Polym. Sci. Polym. Lett. Ed. 12, 723 (1974).
[68] (a) *W. Jaeger, U. Schulke, K. Dietrich* and *G. Reinisch,* U.K. Patent 1, 311, 526 (1970);
(b) Titangesellschaft Neth. Patent 6,610,998 (1967).
[69] *A. P. Pivovarov, Yu. V. Lukovnikov* and *A. F. Gag,* Vysokomol. Soedin. Ser. A, 13, 2110 (1971).
[70] *V. L. Malinskaya* and *L. M. Bajder,* Khim. Vys. Energ. 3, 91 (1969).
[71] *L. M. Bajder, M. V. Voevodskaya* and *N. V. Fok,* Khim. Vys. Energ. 5, 422 (1971).
[72] *T. Takeshita, T. Tsuji* and *T. Seiki,* J. Polym. Sci. A-1, 10, 2315 (1972); Rep. Progr. Polym. Phys. Japan 15, 563 (1972).
[73] *D. G. Marsh,* Int. Symp. Degradation and Stabilization of Polymers, Brussels (1974), Paper 28.
[74] *L. J. Taylor* and *J. W. Tobias,* J. Appl. Polym. Sci. 21, 1273 (1977).
[75] *N. S. Allen* and *J. F. McKellar,* Brit. Polym. J. 9, 302 (1977).
[76] *G. R. Lappin,* in *H. F. Mark, N. G. Gaylord* and *N. Bikales* (eds.), "Encyclopedia of Polymer Science and Technology, Vol. 14, p. 125, Interscience, New York (1971).
[77] *A. Adamczyk* and *F. Wilkinson,* J. Appl. Polym. Sci. 18, 1225 (1974).
[78] *G. A. George,* J. Appl. Polym. Sci. 18, 117 (1974).
[79] *D. J. Carlsson* and *D. M. Wiles,* J. Polym. Sci. Polym. Chem. Ed. 12, 2217 (1974).
[80] *N. S. Allen* and *J. F. McKellar,* J. Appl. Polym. Sci. 22, 3277 (1978).
[81] *R. A. R. Patel* and *J. J. Usilton,* "Ultraviolet Stabilization of Polymers: Development with Hindered-Amine Light Stabilizers" in *D. L. Allara* and *W. L. Hawkins* (eds.), "Stabilization and Degradation of Polymers", Adv. Chem. Series 169, Am. Chem. Soc., Washington (1978), p. 116.
[82] *C. S. Foote* and *S. Wexler,* J. Am. Chem. Soc. 86, 3879 (1964).
[83] *D. J. Carlsson* and *D. M. Wiles,* J. Polym. Sci., Polym. Lett. Ed. 14, 493 (1976).
[84] *M. L. Kaplan* and *P. G. Kelleher,* J. Polym. Sci. A 18, 3163 (1970).
[85] *A. Zweig* and *W. A. Henderson,* J. Polym. Sci. Polym. Chem. Ed. 13, 993 (1975).
[86] *E. F. J. Duynstee* and *M. E. A. H. Mavis,* Eur. Polym. J. 8, 1375 (1972).
[87] *P. Bortolus, S. Dellonte, G. Beggiato* and *W. Corio,* Eur. Polym. J. 13, 185 (1977).
[88] *A. M. Trozzolo* and *F. H. Winslow,* Macromolecules 1, 98 (1968).
[89] *J. R. MacCallum,* "Reactions of Singlet Oxygen with Polymers", in *N. Grassie* (ed.), "Developments in Polymer Degradation-1", Appl. Science Publ., London (1977).
[90] (a) *J. F. Rabek* and *B. Rånby* (eds.), "Singlet Oxygen: Reactions with Organic Compounds and Polymers", Wiley, New York (1978);
(b) "Singlet Molecular Oxygen", Proc. Symp. at Bhabha Atomic Research Centre, Bombay, 1975, INSDOC, Delhi (1978).
[91] *D. R. Kearns,* Chem. Rev. 71, 395 (1971).
[92] *J. C. W. Chien,* "Hydroperoxides in Degradation and Stabilization of Polymers", in [5], p. 95.
[93] *V. S. Pudov* and *A. L. Buchachenko,* Uspekhi Khim. 39, 130 (1970) [Russ. Chem. Rev. 39, 70 (1970)].
[94] *G. Geuskens, D. Baeyens-Volant, G. Delaunois, Q. Lu-Vinh, W. Piret* and *C. David,* Eur. Polym. J. 14, 291 (1978).
[95] *J. E. Guillet,* "Polymers with Controlled Lifetimes", in *J. E. Guillet* (ed.), "Polymers and Ecological Problems", Plenum Press, New York (1973).

[96] G. Scott, "Delayed Action Photo-Activator for the Degradation of Packaging Polymers" in J. E. Guillet (ed.), "Polymers and Ecological Problems", Plenum Press, New York (1973).

[97] M. U. Amin and G. Scott, Eur. Polym. J. 10, 1019 (1974).

[98] (a) W. S. de Forest, "Photoresist, Materials and Processes", McGraw-Hill, New York (1975);
    (b) A. F. Bogenschütz (ed.), "Fotolacktechnik", Leuze-Verlag, Saulgau (1975);
    (c) J. Kosar, "Light-Sensitive Systems", Wiley, New York (1965).

[99] M. J. Bowden and L. F. Thompson, Solid State Technol. 22, 81 (1979).

[100] M. Tsuda, S. Oikawa, Y. Nakamura, H. Nagara, Y. Yokota, H. Nakane, T. Tsumori, K. Nakane and T. Mifune, Int. Symp. Adv. Photopol. Syst., Washington (1978), Paper Summaries, p. 77.

[101] for a survey of negative resists see: E. D. Feit in S. P. Pappas (ed.), "U.V. Curing: Science and Technology", Technology Marketing Corp. Stamford, Con. (1978).

[102] M. Harita, M. Ishikawa, K. Harada and T. Tsunoda, Polym. Eng. Sci. 17, 372 (1977).

[103] M. J. Bowden and E. A. Chandross, J. Electrochem. Soc. 122, 1370 (1975).

[104] C. W. Wilkins, E. D. Feit and M. E. Wurts, Proc. 8th Intern. Conf. Electreon and Ion Beam Sci. and Technol., Electrochem. Society Meeting, Seattle, May (1978).

[105] see e.g.: L. Pacts, E. E. Hardy and M. L. Rodenburg, Ind. Eng. Prod. Res. Develop. 9, 21 (1970).

[106] R. M. Lum, J. Appl. Polym. Sci. 20, 1635 (1976).

[107] J. F. Ireland and P. A. H. Wyatt, J. Chem. Soc. Faraday Trans. I, 68, 1053 (1972); 69, 161 (1973).

[108] A. Weller, Z. Elektrochem. 60, 1144 (1956).

[109] W. Klöppfer, J. Polym. Sci. C 57, 205 (1976).

[110] T. Werner and H. E. A. Kramer, Eur. Polym. J. 13, 501 (1977); T. Werner, J. Phys. Chem. 83, 320 (1979).

5 Degradation by High Energy Radiation

5.1 Introduction

5.1.1 Types of Radiation and their Relevance to Polymer Degradation

The importance of the effects of high energy radiation on polymers is not so obvious as that of the other modes of degradation, dealt with in this book. Therefore, it seems to be appropriate to start with a brief description of the types of radiation in conjunction with relevant applications in the field of polymer degradation.

The term high energy (or ionizing) radiation comprises all kinds of electromagnetic or corpuscular (particle) radiation having quantum or kinetic energies appreciably higher than bond dissociation energies. The modes of electromagnetic radiation, dealt with in this chapter, refer to γ-rays and X-rays. γ-Rays are generated by nuclear reactions, which frequently proceed very rapidly. There is a number of "slow" reactions which are interesting since the respective nuclei can be used as γ-ray sources. Convenient γ-sources are the "long-lived" radioactive nuclei ^{60}Co and ^{137}Cs having halflife times of 5.26 years and 30.2 years respectively[*]). ^{60}Co emits two photons of 1.1 and 1.3 MeV, and ^{137}Cs emits 0.6 MeV photons.

X-rays are produced by allowing accelerated electrons to strike appropriate targets. Thereby, a *continuous spectrum* of bremsstrahlung is produced as a result of the interaction of the atomic nuclei of the target with approaching fast electrons. Moreover, *characteristic radiation* is generated, if an electron of one of the innermost shells is expelled by the impact of a fast electron and the missing electron is replaced by an electron from an outer orbit, the energy of the emitted photon corresponding to the energy difference between the two orbits. Table 5.1 lists target materials and respective wavelengths of characteristic emission lines. Recently "soft" X-rays ($\lambda > 4$ Å) have attracted great attention because of their applications in polymer degradation in conjunction with X-ray lithography and X-ray resists. As powerful sources producing parallel beams are desirable for these purposes, it appears that *synchrotron radiation* is most appropriate. Synchrotron radiation sources are by far the brightest sources of soft X-rays. Synchrotron radiation is incoherent electromagnetic radiation emitted by high energy relativistic electrons circulating in a synchrotron or storage ring, i.e. by electrons which are accelerated normal to the direction of their motion by a magnetic field. The spectrum extends from the microwave into the X-ray region. The wavelength range corresponding to the greatest spectral output is determined by the "characteristic" or "critical" wavelength λ_c (in Å):

$$\lambda_c = 5.6 \frac{R_M}{E^3} = \frac{187}{HE^2} \tag{5.1}$$

R_M: "magnetic" radius (m), at the point of light emission
E: kinetic energy of electrons (GeV)
H: magnetic field (kOe)

---

*) U. Reus, W. Westmeier and I. Warnecke, "Gamma-Ray-Catalog" GSI Report 79-2

Table 5.1   Conventional X-ray sources: characteristic emission lines *)

Target	Series	Wavelength of characteristic emission line (Å) **)
Cu	K_α	1.54
Cr	K_α	2.3
Pd	L_α	4.37
Rh	L_α	4.6
Al	L_α	8.3
Cu	L_α	13.3
C	K_α	44.8

*) adopted from *E. F. Kaelble* (ed.), "Handbook of *X*-Rays", McGraw-Hill, New York (1967)
**) 10 Å = 1 nm

Fig. 5.1 (a) describes the geometry of synchrotron radiation emission and a typical emission spectrum is shown in Fig. 5.1 (b). For literature on modern conventional *X*-ray sources and on synchrotron radiation the reader is referred to [1−3].

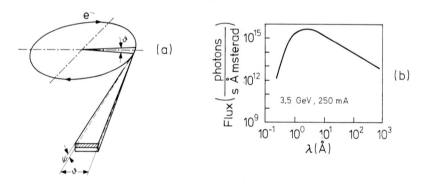

Fig. 5.1   Schematic diagram of generation of synchrotron radiation.
(a) Emitting electron in orbit. ψ = angular spread of beam in vertical direction; ϑ = angular spread in horizontal direction corresponding to the total angular change of direction of the electron within the magnetic field.
(b) Spectrum of electromagnetic radiation emitted at an electron energy of 3.5 GeV (1 GeV: 10^9 eV), sterad $\equiv$ steradian, unit of solid angle; 1 sterad: solid angle which encloses a surface on the sphere equivalent to the square of the radius. (Obtained at Deutsches Elektronen-Synchrotron (DESY), Hamburg [1])

With regard to radiation chemical and nuclear chemical applications during the past decades, the most prominent modes of particle radiation have been fast electrons as well as fast and slow (thermal) neutrons. Fast electron beams are generated in Van de Graaff machines or linear accelerators. Typical specifications are: electron energies from 0.5 to 35 MeV and a beam current of several mA in a continuous operation mode. Neutrons are produced in nuclear reactors (highest flux of thermal neutrons: $10^{15} cm^{-2} s^{-1}$). There is a broad spectrum of particle radiations available at present, ranging from light to heavy nuclei of various charge state (e.g. H, He, Ar, Kr, Xe, Pb and U). They can be routinely accelerated to energies of several 10^8 eV, but because of their very short penetration depth in condensed matter, they are only of academic interest as far as polymers and other organic compounds are concerned.

γ-Rays, X-rays, neutron- and electron-radiation are used for various technical and medical purposes. Since it is common practice to use plastic parts and articles in all sorts of technical equipment, problems of polymer radiation stability are ubiquitous in radiation applications. The radiation resistance of polymers also became an important issue recently, when it was proposed that plastic articles used for medical purposes, such as implants, should be sterilized by irradiation with γ-rays. Radiation sterilization is a convenient method which is utilized to a still increasing extent in various countries. Moreover, problems concerning the radiation stability of polymers usually arise when plastic articles (insulations, glues etc.) are applied in places where high energy radiation prevails, for example, in nuclear power plants or in cancer radiation treatment centers.

It should be emphasized that the field of radiation chemistry of polymers comprises both synthetic polymers and biopolymers. Actually, the behavior of many biopolymers, such as proteins and DNA under the influence of high energy radiation has been studied intensely during the past decades.

There exist a number of interesting technical applications of radiation processes involving polymers, such as the utilization of certain polymers as electron resists, which will also be briefly described in this chapter. A related field of radiation processing, *plasma chemistry*, has become an established science and some applications of plasma reactions in the field of polymers will be discussed in the final section of this chapter. Many articles and books concerning the radiation chemistry of polymers have been published during the last two decades and the reader is referred to some of them for further information [1–12].

5.1.2   Absorption of Radiation

Contrary to the absorption of UV and visible light, dealt with in Chapter 4, high energy radiation is absorbed non-specifically, i.e. there are no chromophores for γ-rays or fast electrons. The absorption of high energy electromagnetic radiation and of fast particle radiation by matter occurs via interactions with both the nuclei of atoms and the clouds of electrons surrounding them.

Interactions of the incident radiation with atomic nuclei can be neglected if the photon or kinetic energies, respectively, are lower than about 10 MeV and if the irradiated material consists only of light nuclei. With organic polymers, consisting essentially of C, O, H, N, S, P, this is, actually, the case.

If γ-rays or X-rays interact with electrons in atomic or molecular orbitals, three processes are possible: photoelectric effect, Compton effect, and pair formation, their relative importance depending on photon energy and atomic number of the nuclei as well as on the electron density of the irradiated system. In all three cases secondary electrons are ejected, which usually possess sufficient kinetic energy to induce additional ionizations or electronic excitations in surrounding molecules:

$$M \xrightarrow{\text{high energy photon}} M^{+\bullet} \quad + \quad e^-_{kin} \tag{5.2}$$

radical   secondary
cation   electron

$$M \quad e^-_{kin} \Bigg\langle \begin{array}{l} \longrightarrow \quad M^{+\bullet} \quad + \quad e^-_{kin} \tag{5.3a} \\[2em] \longrightarrow \quad M^* \tag{5.3b} \end{array}$$

electronically
excited molecule

Reaction (5.2) describes primary ionizations, whereas the interaction of secondary electrons with intact molecules is represented by (5.3 a) and (5.3 b).

It is interesting to note that reactions (5.3 a) and (5.3 b) also apply for the absorption of corpuscular radiation. Principally, fast electrons lose their energy by inelastic and elastic collisions as well as by the generation of bremsstrahlung. According to the restrictions previously mentioned elastic collisions and the generation of bremsstrahlung can be neglected, and so only inelastic collisions

$$e^-_{fast} \quad + \quad M \quad \longrightarrow \quad M^{+\bullet} \quad + \quad e^-_{kin} \quad + \quad e^-_{fast} \tag{5.4}$$

primary   radical   secondary   primary
electron   cation   electron   electron

(E_{kin}) (E'_{kin})

have to be considered ($E'_{kin} < E_{kin}$).

It must be pointed out that the kinetic energy of most secondary electrons is less than 100 eV, which implies that they lose their energy in close proximity to their origin. On the other hand, fast electrons, of say, $E_{kin} = 1$ MeV, pass many molecules on their way without interaction. Only occasionally do they interact with electrons located in molecular or atomic orbitals. Therefore, it can be concluded that, in these cases, the absorption of energy occurs rather heterogeneously. According to the terminology used by radiation chemists, a "track" of a high energy electron (e.g. 1 MeV) consists of a series of "spurs" having an average distance of a few thousand Å. Most of the spurs contain less than four ion pairs and a corresponding number of excited molecules. The distribution of spurs is subject to the "linear energy transfer" (LET) of a particle traversing matter, i.e. the amount of energy dissipated per unit path length. The LET increases, as the atomic number of the stopping material, the particle mass and the charge of the particle increase. With increasing LET, the average distance between spurs is reduced. Actually overlapping of spurs is a general phenomenon in irradiations with charged high energy particles, such as protons, α-particles or products of nuclear fissions. If low LET radiation, e.g. γ-rays and fast electrons are absorbed, a great portion of intermediates formed in the spurs is able to diffuse out and to react with solutes or species generated in other spurs.

For more detailed information the reader is referred to the relevant literature [16—22].

5.2 Mechanistic Aspects

5.2.1 Reactive Intermediates

In the course of the interaction of ionizing radiation with matter, ions, radicals and electronically excited molecules are generated. All these species are unstable, i.e. they are more or less reactive towards intact molecules. In Scheme 5.1 various important elementary reactions are compiled with the intention to show how transient species are generated, or converted to longer-lived intermediates. The formation of *stable products* in a simple decay process, as exemplified by reaction (e) in Scheme 5.1, is a very rare event. The overwhelming portion of stable products are formed via radical-radical or ion-ion (neutralization) processes.

Scheme 5.1   Generation and decay of intermediates

Primary radiolytic act	$M \longrightarrow M^{+\cdot} + e_{kin}^-$	(a)
Capture of thermalized electron by positive ion	$M^{+\cdot} + e_{th}^- \rightarrow M^*$	(b)
Capture of thermalized electron by neutral molecule	$M + e_{th}^- \rightarrow M^{-\cdot}$	(c)
Bond breaking reactions:		
Decomposition of excited molecule (radical formation)	$M^* \rightarrow R_1^\cdot + R_2^\cdot$	(d)
Decomposition of excited molecule (formation of stable products)	$M^* \rightarrow A + B$	(e)
Decomposition of positive radical ion	$M^{+\cdot} \rightarrow R^\cdot + C^+$	(f)
Decomposition of negative radical ion	$M^{-\cdot} \rightarrow R^\cdot + D^-$	(g)
Side reactions:		
Charge transfer	$M^{+\cdot} + S \rightarrow M + S^{+\cdot}$	(h)
	$M^{-\cdot} + S \rightarrow M + S^{-\cdot}$	(i)
Energy transfer	$M^* + S \rightarrow M + S^*$	(j)

Evidence for transient species has been obtained by techniques similar to those used in photodegradation studies (see Chapter 4). Apart from stationary irradiations at low temperatures, the pulse radiolysis technique has proved useful to detect transient species and to follow the kinetics of their reactions [25].

The electron spin resonance method is a very valuable tool to detect and identify *free macroradicals* [26−30], which in certain rigid systems, i.e. at temperatures below T_g, have lifetimes of several weeks or months. With the aid of UV absorption spectroscopy, reactions of macroradicals can be followed in transparent systems.

Electronically excited states have also been demonstrated by optical absorption measurements. In several cases, appropriate low molecular weight additives served as traps for excited states. Triplet excited naphthalene and biphenyl molecules were formed in matrices of polystyrene and polymethylmethacrylate upon electron irradiation [31]. Owing to their very high extinction coefficients, these triplets are readily detectable. Since direct excitation could be excluded, it was inferred from these results that excited states are primarily formed in the polymer with subsequent excitation energy transfer to the additive.

Electrically charged species (ions) have been demonstrated by optical absorption and emission [31—33] as well as by electrical conductivity [34, 35] measurements. In investigations with rigid polymers, using optical detection methods, low molecular weight additives mostly served as traps. The formation of radical cations and anions of the additive, which was present at low concentrations, indicated that charge transfer processes proceed readily. The existence of macroions has been demonstrated directly only in a few cases, e.g. with polystyrene, where negative ions were detected [31]. Electrical conductivity measurements in conjunction with pulse radiolysis of rigid polymers revealed two modes of conduction, as shown in Fig. 5.2 [35]. In aromatic polymers, such as polyethyleneterephthalate, polyvinylnaphthalene and polystyrene, the rapid mode was correlated to a conduction mechanism involving the π-electron systems of the aromatic rings, whereas the slow mode was considered to be due to carriers migrating through the system via repeated trapping and untrapping processes. Phenomenologically similar results, obtained in polyethylene, were explained on the basis of electrons traveling through crystalline and amorphous regions of the polymer specimen. The crystalline regions, possessing only energetically "shallow" traps, permit rapid migration of electrons, the amorphous regions, on the other hand, possessing mainly "deep" traps, impede carrier migration.

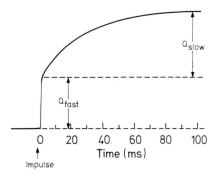

Fig. 5.2 Evidence for the radiation-induced generation of charge carriers in polyethylene. Schematic illustration of charge collection at Au electrodes attached to a polyethylene foil, after irradiation with an impulse of 10 keV electrons. (According to [35])

Because in many cases transient species are rather short-lived and processes frequently occur according to several reaction routes, our knowledge about mechanisms of high energy radiation-induced degradation processes in polymers is still limited and further elucidation is highly desirable. Whereas a more detailed discussion of mechanistic aspects would go beyond the scope of this chapter, the mechanism of main-chain degradation of polymethylmethacrylate, presented in Scheme 5.2, gives a general impression of how radiation-induced reactions in polymers proceed.

Scheme 5.2.   Mechanism of main-chain scission in polymethylmethacrylate [36]

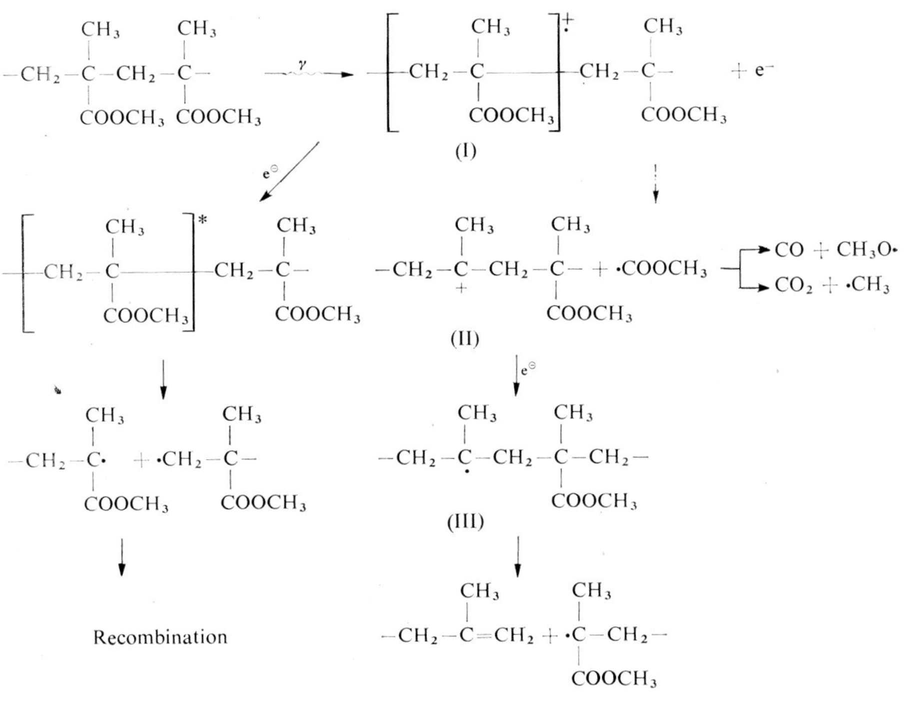

$$
\begin{array}{c}
\overset{\displaystyle CH_3}{\underset{\displaystyle COOCH_3}{|}}\quad \overset{\displaystyle CH_3}{\underset{\displaystyle COOCH_3}{|}} \\
-CH_2-\underset{|}{C}-CH_2-\underset{|}{C}- \xrightarrow{\;\gamma\;}
\left[-CH_2-\underset{\underset{COOCH_3}{|}}{C}- \right]^{\!\!+\!\cdot}
\; \underset{\underset{COOCH_3}{|}}{\overset{\overset{CH_3}{|}}{CH_2-C}}- \; + e^-
\end{array}
$$

(I)

$e^{\ominus}$

$$
\left[-CH_2-\underset{\underset{COOCH_3}{|}}{\overset{\overset{CH_3}{|}}{C}}- \right]^{\!*}
\; -CH_2-\underset{\underset{COOCH_3}{|}}{\overset{\overset{CH_3}{|}}{C}}-
$$

$$
-CH_2-\underset{|}{\overset{\overset{CH_3}{|}}{C}}-CH_2-\underset{\underset{COOCH_3}{|}}{\overset{\overset{CH_3}{|}}{C}}- \; + \; \cdot COOCH_3
\left\{
\begin{array}{l}
\rightarrow CO + CH_3O\cdot \\
\rightarrow CO_2 + \cdot CH_3
\end{array}
\right.
$$

(II)

$e^{\ominus}$

$$
-CH_2-\underset{\underset{COOCH_3}{|}}{\overset{\overset{CH_3}{|}}{C}}\cdot \; + \; \cdot CH_2-\underset{\underset{COOCH_3}{|}}{\overset{\overset{CH_3}{|}}{C}}-
\qquad
-CH_2-\underset{\underset{\cdot}{}}{\overset{\overset{CH_3}{|}}{C}}-CH_2-\underset{\underset{COOCH_3}{|}}{\overset{\overset{CH_3}{|}}{C}}-CH_2-
$$

(III)

Recombination

$$
-CH_2-\overset{\overset{CH_3}{|}}{C}=CH_2 \; + \; \cdot \overset{\overset{CH_3}{|}}{\underset{\underset{COOCH_3}{|}}{C}}-CH_2-
$$

5.2.2   Simultaneous Crosslinking and Degradation

Except for a few cases (e.g. polymethylmethacrylate and polytetrafluoroethylene) where main-chain scission is the exclusive process altering the molecular size, linear polymers undergo both crosslinking and main-chain scission when exposed to high energy radiation. As a rule, polymers with tetrasubstituted carbons in the repeating unit are predominantly ruptured in the main-chains, as indicated by a decrease in molecular weight with increasing absorbed dose. Poly(phenylvinyl ketone) is an exeption, because it undergoes predominantly main-chain scission, inspite of having no tetrasubstituted carbons in the backbone. However, this polymer probably undergoes also, type II scission when subjected to high energy irradiation, in analogy to its behavior on irradiation with UV light.

The determination of radiation chemical yields*) of main-chain scissions $G(S)$ and crosslinks $G(X)$ is relatively easy, if the initial MWD is of the "most probable" type (see Section 1.4). In this case the number and weight average degrees of polymerization u_1 and u_2, respectively, depend on the absorbed dose D (in eV/g) according to Eq. (5.5) and (5.6):

$$\frac{1}{u_1} = \frac{1}{u_{1,0}} + (G(S) - G(X))\frac{Dm}{100N_A} \tag{5.5}$$

$$\frac{1}{u_2} = \frac{1}{u_{2,0}} + \left(\frac{G(S)}{2} - 2G(X)\right)\frac{Dm}{100N_A} \tag{5.6}$$

N_A = Avogadro's number: m = molecular weight of structural repeating unit

It should be noted, that only Eq. (5.5) holds for any initial MWD. Three typical cases of MW changes as a function of absorbed dose, can be visualized, depending on whether $G(S)/G(X)$ is equal to, greater than or smaller than 4. This is illustrated in Fig. 5.3.

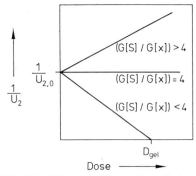

Fig. 5.3 Schematic illustration of the variation of the weight average degree of polymerization with absorbed dose. Plots of u_2^{-1} vs. dose at various values of G(S)/G(X)

$G(S)$ and $G(X)$ can be determined by measuring both u_2 and u_1 as a function of the absorbed dose. If crosslinking predominates, the polymer becomes insoluble at $D > D_{gel}$ and $G(S)$ and $G(X)$ can be determined with the aid of the Charlesby-Pinner equation:

$$s + s^{1/2} = \frac{G(S)}{2G(X)} + \frac{100N_A}{u_{2,0}\, G(X)\, mD} \tag{5.7}$$

where "s" denotes the insoluble fraction. Equation (5.7) holds only for simultaneous "random" main-chain scission and crosslinking and for a "most probable" initial MWD.

$G(S)$ and $G(X)$ values presented in Table 5.2 were determined with the aid of the methods described above.

---

*) G-values denote the number of molecules or atoms produced or decomposed per 100 eV
   of absorbed energy. $G(S)$ and $G(X)$ designate numbers of broken main-chain bonds and
   newly formed intermolecular linkages per 100 eV, respectively.

Table 5.2   Radiation chemical yields of main-chain scission and intermolecular crosslinking of various polymers, irradiated with low LET radiation at room temperature in vacuo or in the presence of an inert gas (from ref. [15])

Polymer	$G(S)$	$G(X)$	Dominating Process *)
polyethylene		2.0	crosslinking
polyisobutene	1.5−5.0	<0.05	scission
polystyrene	0.02	0.03	crosslinking
poly-α-methylstyrene	0.25		scission
polymethylmethacrylate	1.2−2.6		scission
polytetrafluoroethylene	0.1−0.2		scission
poly(ethylene oxide)	2.0	1.8	crosslinking
polyphenylvinylketone	0.35		scission
poly(propylene sulfide)	0.4−0.7		scission
polydimethylsiloxane	0.07	2.3	crosslinking
poly(butene-1 sulfone)	12.2		scission
poly(hexene-1 sulfone)	10.7		scission
amylose	1.3−2.7		scission
cellulose	3.3−6.8		scission
polylysine	4.1		scission
polyalanine	3		scission
DNA	0.8		scission

*) scission denotes predominant main-chain scission; crosslinking denotes predominant intermolecular crosslinking (gel forming type)

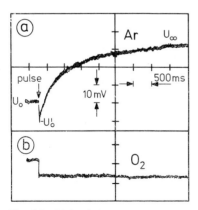

Fig. 5.4   Main-chain scission and intermolecular crosslinking in poly(methylvinyl ketone) induced by a 2 μs pulse of 15 MeV electrons (2×10^4 rad), as determined by light scattering (LS) measurements. The oscilloscope traces, obtained at room temperature with acetone solutions (20 g/l) saturated with argon (a) or with O_2 (b), show the LS change after the pulse [51]

Kinetic investigations in the author's laboratory [50] revealed, that in several cases "simultaneous" main-chain rupture and intermolecular crosslinking can be readily distinguished with the aid of time-resolved Rayleigh light scattering measurements. In the case depicted in Fig. 5.4 polymethylvinylketone was irradiated in acetone solution with a 2 μs pulse of 15 MeV electrons [51]. Immediately after the pulse the light scattering intensity decreased, indicating a decrease of the MW and thus main-chain scission. At a later stage, the light scattering intensity increased indicating intermolecular crosslinking. Molecular oxygen decreased the extent of main-chain scission and completely prevented crosslinking, as depicted in Fig. 5.4 (b).

5.3   Special Aspects of Degradation in Bulk Synthetic Polymers

5.3.1   Influence of Oxygen

High energy radiation is an excellent means to initiate autoxidation, because, as was outlined above, upon irradiation of organic polymers free radicals are formed as intermediates. For more information the reader is referred to Section 1.3.2, where autoxidation has been treated in some detail. Let us only recall here that main-chain scission can occur in the course of reactions involving peroxyl and oxyl radicals. As a consequence, it can be anticipated that in the presence of O_2 and at low absorbed dose rates, $G(S)$ will be higher than in its absence. (Dose rate effects are discussed in the next section.) Moreover, O_2 is expected to react with lateral macroradicals

$$\sim\!\!\sim\!\!\sim \;\;+\;\; O_2 \;\;\rightarrow\;\; \sim\!\!\sim\!\!\sim \qquad\qquad (5.8)$$
$$\underset{O-O\bullet}{\mid}$$

thus preventing crosslinking. Because, at low absorbed doses, crosslinking frequently improves the mechanical properties of polymers, its prevention together with the increased main-chain scission yield, leads to a rapid deterioration and subsequent breakdown of mechanical properties. Therefore, high energy radiation in conjunction with oxygen affects most polymers in a highly deleterious manner. Many polymers of the crosslinking type, such as polyethylene [37, 38], polypropylene [7], polystyrene [38—40] and polyvinylchloride [41, 42], undergo predominantly main-chain scission upon irradiation in air. Moreover, polytetrafluoroethylene undergoes main-chain rupture in oxygen or air at a much higher yield than in vacuo [43]. There are, however, some polymers, which are less affected by radiation if O_2 is present. A representative of this group is polymethylmethacrylate. Though this effect is not fully understood, it is very likely that oxygen acts as quencher for certain electronically excited repulsive states reached in polymethylmethacrylate under irradiation.

5.3.2   Influence of Absorbed Dose Rate

As far as low LET radiation is concerned, true dose rate effects in polymers have not been demonstrated. Upon subjecting polyethylene to fast electron irradiation in vacuo, for example, $G(X)$ was independent of the absorbed dose rate in the broad range from 0.1 to 34.5 Mrad/s [44]. Analogously, $G(S)$ remained constant in the case of polymethylmethacrylate when the dose rate was varied from 0.5 Mrad/h to several Mrad/h [45].

Quite frequently, however, apparent dose rate effects are encountered, the most prominent one relying on the consumption of oxygen in oxidative degradation. If the dose rate is rather high, it can happen that after a certain irradiation time the O_2 concentration in the interior of the specimen is depleted. Therefore, the outer layer can predominantly undergo main-chain scission, while deeper layers become insoluble because of cross-linking. This phenomenon has been demonstrated recently with polystyrene [40].

Another source of apparent dose rate effects refers to an increase in the specimen's temperature. Since the major portion of the absorbed energy is converted to heat, at high dose rates the heat release to the environment can be insufficient resulting in an increase in the specimen's temperature. As many chemical reactions have a rather high activation energy, it is feasible that at high absorbed dose rates, i.e. at high temperatures, reaction pathways different from those occurring at low dose rate, when the specimen's temperature remains unchanged during irradiation, can prevail.

5.3.3   Influence of LET

Radiation chemical studies on LET effects in low molecular weight compounds have shown that so-called "molecular product" yields increase with increasing LET. Molecular products are thought to be generated in the spurs, before the latter expand and enable the reactive species to diffuse into the bulk of the system. On the basis of this definition (molecular products = products of intraspur reactions), it becomes feasible that spur overlapping, as encountered with high LET radiation, increases the probability of molecular product formation. If this idea is applied to polymers, multiple ionizations and excitations in proximate repeating units of the chains can be anticipated, with the consequence that the products formed will differ qualitatively and quantitatively from those in low LET irradiations. Unfortunately, up to now, LET effects in polymers have been investigated only occasionally. A compilation of relevant work can be found in the book of Makhlis [14]. With polymethylmethacrylate, $G(S)$ was found to decrease with increasing LET. Furthermore, $G(X)$ was reported to be independent of LET in the cases of polyvinylchloride and polyethylene, whereas polystyrene was found to be cross-linked more efficiently by high LET-particles than by sparsely ionizing radiation. The results reported in the literature are, to some extent, at variance. It is hoped, therefore, that as heavy ion accelerators become more available to radiation chemists, this interesting field will be investigated more thoroughly.

5.3.4   Temperature Effects

Three aspects should be pointed out: (I) High energy radiation is very well suited to initiate thermal depolymerization (unzipping) of polymers prone to this type of reaction, the most prominent examples being poly(olefin sulfones). For poly(butene-1 sulfone) it has been reported [46] that at 30 °C $G(S) = 12.2$ and G(total gas) $= 39$, the latter value increasing to 478 at 72 °C. As indicated by the rather high value of 39 at 30 °C, unzipping occurs already at ambient temperature. Because of their pronounced tendency to unzip, several polysulfones are very suitable for use as electron and X-ray resists (see Section 5.5.1). (II) Rates and product yields of radiation-induced chemical reactions in many polymers are correlated with molecular mobility. This explains why increases in temperature, leading to phase transitions or allowing specific intramolecular motions (rotations of side groups etc.), frequently influence radiation chemical yields in polymers. Increasing the temperature generally reduces the probability of cage reactions (e.g., radical recombinations). As a typical example, the temperature dependence of $G(S)$ of polyisobutene is shown by an Arrhenius plot in Fig. 5.5 [47]. (III) There is no general correlation between thermal stability and radiation resistance. For instance, under the influence of high energy radiation, polytetrafluoroethylene readily undergoes main-chain scission and polysiloxanes are efficiently crosslinked, although both polymers are rather heat resistant. On the other hand, various polymers, such as the polyimides combine good heat stability with high resistance against ionizing radiation.

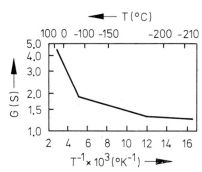

Fig. 5.5   Temperature dependence of main-chain scission in polyisobutene. Arrhenius plot of G(S) vs. T^{-1}. (According to *Wündrich* [47])

5.3.5   Radiation Stability and Protection

Generally, the mechanical properties of *polymers of the crosslinking type* are improved upon irradiation with relatively low doses (about 10 Mrad, depending on the nature of the polymer and the initial MW): for instance, elastic modulus, tensile strength and hardness increase, whereas the solubility decreases with increasing dose. Therefore, radiation treatment of several polymers is commercially utilized, most frequently with polyethylene in the form of cable insulations or shrinkable foils. However, at high absorbed doses, these polymers becomes very hard and brittle.

On the other hand, the properties of polymers undergoing predominantly main-chain scission deteriorate steadily upon irradiation, even at low absorbed doses.

Oxygen exerts a deleterious influence, which, in many cases, results in an acceleration of the deterioration rate, the latter effect being most pronounced in thin specimens and at low dose rates, as discussed in Sections 5.3.1 and 5.3.2.

Protection of polymeric materials against radiation damage can be achieved to some extent by external shielding. Internal screening is usually not applicable. External screening is relatively easy to accomplish at low penetration depths of the incident radiation, such as in the case of soft X-rays or high LET particle radiation. The necessary thickness for sufficient screening is given by the absorption characteristics of the shielding material and by the photon or particle energy. With highly penetrating radiations, such as hard X-rays or γ-rays, external shielding can only be achieved by rather thick layers of strongly absorbing material, e.g. lead.

Protective agents (antirads), added to the polymer in low concentration, are only effective to a limited extent, as the overwhelming portion of the radiation energy is absorbed by the polymer. Important modes of action of antirads pertain (I) to the prevention of oxidative chain reactions (antioxidant action), or more generally, to their reactions with reactive radicals (radical scavenger action) and (II) to transfer of excitation energy from the polymer to the additive (energy sink action). In many cases aromatic compounds or fillers, consisting of inorganic material or carbon black, exhibit significant protective properties.

Frequently, however, it is difficult or impossible to find antirads giving reasonable and longlasting protection. As a result of this rather unsatisfactory situation, polymers of sufficiently high radiation stability are needed. Polymers with aromatic constituents may be appropriate, because, generally, aromatic compounds are only slightly affected compared to non-aromatic compounds. As a matter of fact, as far as tensile strength, elongation, elastic modulus, shear strength and impact strength are concerned, polystyrene, if irradiated in an inert atmosphere, is not appreciably damaged by absorbed doses of up to 1 000 Mrad. If, however, oxygen is present in the specimen, these mechanical properties are decreased to 50% of their initial value even after a dose of ca. 80 Mrad [40]. An aromatic polysulfone with the structure

proved to be quite resistant in the absence of O_2. No change in flexural yield strength was detected after doses up to 600 Mrad at $35° - 125°C$. However, in the presence of air, this property decreased above 100 Mrad [48]. Among the polymers of highest radiation stability are aromatic polyimides, such as polypyromellitimide with the structure

The mechanical and electrical properties of this polymer are reported to remain satisfactory up to absorbed doses of 10000 Mrad [49].

5.4   Radiation Effects in Biopolymers

Biopolymers, subjected to high energy radiation, undergo chemical changes with the consequence of a partial or a total loss of their biological functionality. Since, quite often, very big single macromolecules function as a whole unit, a chemical change in a small portion of the polymer backbone or in a single side group can be catastrophic. This situation prevails, e.g., in the case of deoxyribonucleic acid (DNA), the MW of which can be extremely high ($10^8 - 10^9$) and the results shown in Table 1.2 apply fully here, i.e. a small chemical conversion implies a large fraction of macromolecules affected. Thus, it becomes comprehensible, why the lethal doses (required to kill 50% of a population within 30 days after exposure) for highly organized species are less than 10^3 rad (for human beings 500 rad and for frogs and chickens 700 rad). The chemical composition of DNA will be described in Section 6.3.1.3. Native DNA consists of double strands (see Fig. 2.16). As shown in Fig. 5.6, single strand breaks can be distinguished from double strand breaks, the latter consisting of pairs of single strand breaks on opposite sites in the double strand. As at room temperature, the critical distance, h_{crit} *), between single strand breaks in paired strands amounts to about $10 - 15$ moieties [53], a great portion of single strand breaks do not result in double strand breaks and can, therefore, become subject to enzymatic repair processes. In accordance with this mechanism are findings obtained on irradiating native DNA in dilute solution, in vitro, which show that the number of double strand breaks is proportional to the square of the absorbed dose [52]. However, upon irradiating DNA in situ, in bacteriophages and in cells, the number of double strand breaks increased linearly with absorbed dose [54], demonstrating, that in vivo mechanisms do not correspond to the mechanism prevailing in

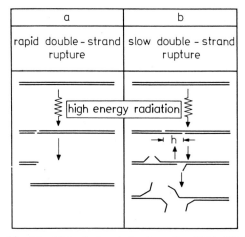

Fig. 5.6   Main-chain scission in native DNA [52, 53].
(a) Single-strand breaks on opposite sites in the double strand. (b) Single-strand breaks separated by h intact moieties. Double-strand scission occurs if $h < h_{crit}$. Pulse radiolysis studies have revealed that fragment separation is more rapid in case (a) than in case (b)

dilute solutions in vitro. The latter mechanism is essentially based on indirect radiation effects, i.e. the radiation-induced chemical changes in DNA are caused by the attack of reactive species stemming from the radiolysis of water. Recent results concerning radiation effects in DNA have been reviewed [55]. For information concerning radiation effects in other biomacromolecules, various books are available [56].

———————

*) at $h > h_{crit}$ single strand breaks do not result in double strand breaks.

5.5   Applications

5.5.1   Electron Beam and X-Ray Lithography

5.5.1.1   Electron and X-ray Resists in Microelectronics

Over the past decade, high energy radiation has been applied successfully in the resist field [1−3, 57]. Principally, there are two modes of action of high energy radiation resists, as in the case of photoresists, i.e. positive and negative action. The reader is, therefore, referred to Section 4.5.2 where the utilization of photoresists was discussed and where also a general description of positive and negative resist techniques was presented (see Fig. 4.12).

So far, electron beams and X-rays or synchrotron-radiation have been used. According to Bowden and Thompson [57], there are two specific areas of application of electron beams: the generation of primary patterns (mask making) and direct writing, in which the circuit configuration is inscribed onto the wafer. With the aid of computer-control, the electron beam can be directed very precisely and reproducibly over the target area.

Table 5.3   Electron Resists (adopted from *Bowden* and *Thompson* [57])

Polymer *)	Mode of Action	Sensitivity D_s**) (Coulomb/cm^2)	Ref.
polymethylmethacrylate	positive	5×10^{-5} (at 15 kV)	57
polymethyl-α-chloro-acrylate	positive	1.2×10^{-5} (at 15 kV)	58
CP-MMA-isobutene (3:1)	positive	5×10^{-6} (at 15 kV)	59
CP-MMA-acrylonitrile (13:1)	positive	4×10^{-6}	60
crosslinked polymeth-acrylamide	positive	$2-3 \times 10^{-7}$	61
crosslinked CP-MMA-methacrylic acid chloride	positive		62, 63
poly(butene-1 sulfone)		8×10^{-7} (at 10 kV)	57, 65
poly(1-methyl-1-cyclo-pentene sulfone)	positive positive	1.3×10^{-6} (at 10 kV)	66
CP-styrene-SO$_2$	positive	10^{-5}	65
CP-glycidylmethacrylate-styrene (4:1)	negative	9.5×10^{-7} (at 10 kV)	68
CP-glycidylmethacrylate-chlorostyrene	negative	$1-2 \times 10^{-6}$ (at 10 kV)	67
epoxy-based polyglycidyl-methacrylates	negative		68−70
epoxidized polybutadiene	negative		68, 69

*) CP stands for copolymer

**) D_s = minimum dose to pattern resist; i.e. dose required to make 100% of the irradiated film soluble (positive resist) or 50% insoluble (negative resist)

In spite of rather impressive results concerning radiation-sensitivity, feature size, developability etc., the search for better resist materials is still under way, with the aim of achieving a minimum feature size of less than one μm.

Some resist materials used and/or tested up to now are presented in Table 5.3.

Crucial requirements for electron resists are high radiation sensitivity and high resolution. Resist properties affecting resolution are contrast, and swelling during development. One of the most popular positive resists is polymethylmethacrylate. Fig. 5.7 (a_1) and (a_2) show electron microscope photographs depicting mask replicas in PMMA obtained with synchrotron radiation. Although these pictures demonstrate that PMMA is quite appropriate for utilization as a positive resist, its radiation sensitivity and its adhesive properties do not fulfill the requirements in many cases. Therefore, various modifications, such as the exchange of the CH_3 group in the α-position versus a chlorine atom, or copolymerization of MMA with other monomers, were investigated. Also of interest are positively acting systems that are chemically crosslinked prior to X-ray or electron irradiation, such as polymethacrylamide:

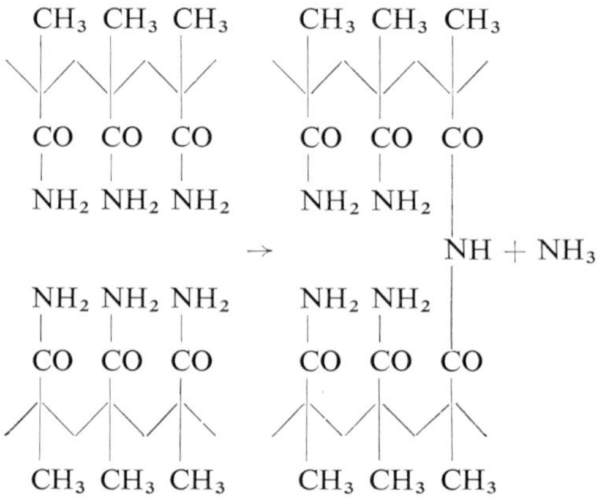

$$
\begin{array}{ll}
CH_3\ CH_3\ CH_3 & CH_3\ CH_3\ CH_3 \\
\quad\ |\quad\ |\quad\ | & \quad\ |\quad\ |\quad\ | \\
CO\ \ CO\ \ CO & CO\ \ CO\ \ CO \\
\ |\quad\ |\quad\ | & \ |\quad\ |\quad\ | \\
NH_2\ NH_2\ NH_2 & NH_2\ NH_2 \\
 & \qquad\qquad\ \ NH + NH_3 \qquad\qquad (5.9)\\
NH_2\ NH_2\ NH_2 & NH_2\ NH_2 \\
\ |\quad\ |\quad\ | & \ |\quad\ |\quad\ | \\
CO\ \ CO\ \ CO & CO\ \ CO\ \ CO \\
\quad\ |\quad\ |\quad\ | & \quad\ |\quad\ |\quad\ | \\
CH_3\ CH_3\ CH_3 & CH_3\ CH_3\ CH_3
\end{array}
$$

and copolymers of MMA and methacrylyl chloride, containing only a small portion of the chloride:

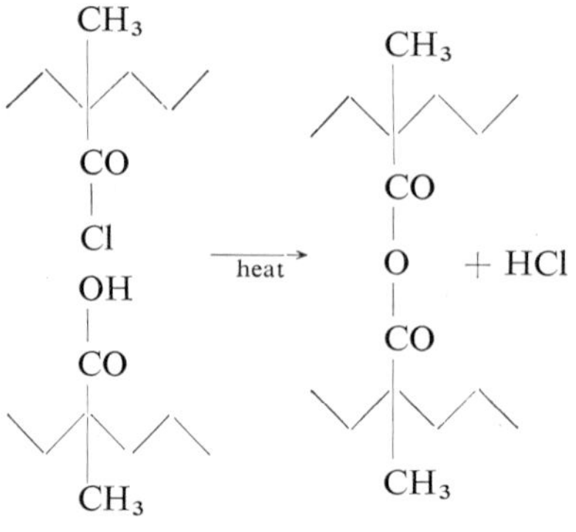

$$
\begin{array}{ll}
CH_3 & CH_3 \\
\ | & \ | \\
CO & CO \\
\ | & \ | \\
Cl & O \qquad + HCl \qquad\qquad (5.10)\\
OH \quad \xrightarrow{heat} & \ | \\
\ | & CO \\
CO & \ | \\
\ | & CH_3 \\
CH_3 &
\end{array}
$$

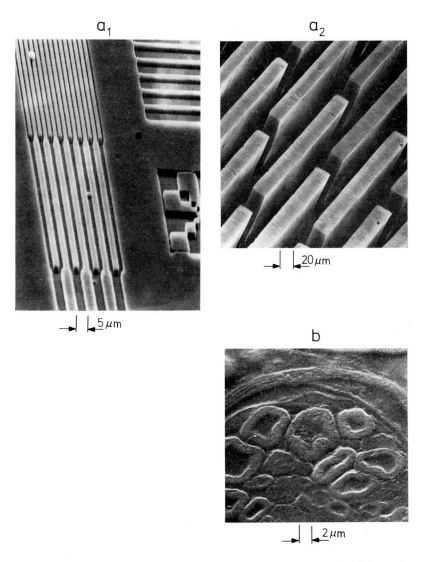

Fig. 5.7 Application of polymer degradation to high energy radiation lithography using synchroton radiation.

(a) Mask replicas of a polymethylmethacrylate resist upon irradiation at DORIS (DESY, Hamburg), (a_1) 60 min irradiated at 1.5 GeV and 30 mA; (a_2) 60 min irradiated at 3.3 GeV and 13.8 mA, dimensions of patterns: 20 μm (width) and 160 μm (depth). (Courtesy of Dr. *A. Heuberger*, Fraunhofer-Institut für Festkörpertechnologie, München).

(b) Replica of part of a cross-section of nerve tissue of rabbit (branch of nervus ischiadicus) in polymethylmethacrylate, 40 s irradiated at the synchrotron in Bonn at 1.5 GeV and 40 mA. (Courtesy of Prof. *Sotobayashi* and Dr. *Asmussen*, Fritz-Haber-Institut, Berlin)

The carboxyl groups, necessary for the crosslinking reaction, are formed upon bringing the copolymer in contact with moist air, which is usually unavoidable, anyway.
Poly(olefine sulfones) of the structure

(R: 1-butene, cyclopentene, 1-hexene, 2-methyl-pentene-1 etc.) act effectively as positive resists [65]. It is noteworthy, that they can be developed in two modes, namely by conventional solvent development and by vapor development. The latter process corresponds to thermal decomposition (unzipping, see Chapter 2) after radiation-induced main-chain rupture:

$$\text{poly(olefin sulfone)} \rightarrow \text{olefin} + SO_2 \qquad (5.11)$$

The rate of vapor development is essentially determined by the chemical nature of the olefin and the exposure temperature.

Copolymers of glycidylmethacrylates and various monomers such as ethylacrylate, styrene, chlorostyrene have prominence among the negatively acting resists. Moreover, epoxydized systems possess interesting properties and have been commercialized in Japan [66–70].

5.5.1.2   X-Ray Resists in Contact Microscopy

Recently, powerful synchrotron radiation sources have become available, as electron storage rings and synchrotrons were constructed by nuclear physicists. This development has resulted in increased activity in the field of X-ray resists and X-ray lithography. Apart from applications in the fabrication of microelectronic devices, synchrotron radiation in the wavelength range $1-10$ nm (1240 to 124 eV) is very well suited for high resolution X-ray microscopy of biological specimens in the natural state, viz. wet and unstained, because of high total absorption cross sections and the absence of scattering [71, 72]. The specimen is brought in close contact to the resist surface and irradiated. After development, a relief structure is obtained, where the height of the feature in the resist corresponds to the X-ray absorption of the specimen. A magnified picture of the resist relief image is produced by a scanning electron microscope. Fig. 5.7 (b) shows a typical example.

It appears possible to attain a linewidth resolution below 0.1 μm at exposure times of less than 0.1 s. Prominent among the polymers used is again polymethylmethacrylate. Because of the relatively low sensitivity of this polymer, improvement was attempted by choosing materials with better absorption characteristics, for example polyhexafluorobutylmethacrylate [64]. Moreover, copolymers of MMA with thallium or barium salts of methacrylic acid have been used to enhance the absorption of X-rays.

Concluding this section, it should be pointed out, that the application of radiation chemical polymer degradation to the resist field and to lithography in general, has so far yielded very fruitful results. At present, various research groups are continuing investigations in these fields and further progress can be expected in the near future.

5.5.2 Solid Lubricants

Polytetrafluoroethylene powders of MW from 3×10^3 to 2.5×10^5 have excellent properties as lubricating powders, because of their low friction coefficient and the low critical surface tension. Commercially available polytetrafluoroethylene has, however, a much higher average molecular weight. Thus, radiation-induced main-chain cleavage in PTFE appears to be an interesting method for the utilization of PTFE scrap. It is reported that this process has found commercial application but details are still proprietary [74].

5.5.3 Graft and Block Copolymerization

High energy radiation generates free radicals in polymers and is, therefore, a means of initiating reactions involving macroradicals. As shown below, irradiation of polymers, in the presence of vinyl monomers, leads to the formation of graft- and/or block copolymers, depending on the location of the radical sites:

lateral radical   graft copolymer

$$\text{radical} + n\,M \longrightarrow \text{graft copolymer with } M, (M)_{n-2}, M \qquad (5.12)$$

terminal radical   block copolymer

$$\text{terminal radical} + n\,M \longrightarrow \text{—}M\text{—}(M)_{n-2}\text{—}M\cdot \qquad (5.13)$$

If radical formation by main-chain rupture dominates over processes in side groups, block copolymerization will dominate over grafting. Radiation-induced graft and block copolymerization is an elegant method to synthesize new polymers with unique properties.

In this connection the reader is reminded of the discussion of mechanochemical graft and block polymerizations in Section 3.4.5.

There are several commercial applications of radiation-induced graft and block copolymerizations. For further details the reader is referred to the literature [75].

5.6 Plasmachemistry

During the last two decades, plasma techniques have been used to an increasing extent by chemists to initiate chemical reactions. A certain fraction of these activities has been devoted to plasma polymerizations and to the modification of surfaces of conventional polymers with the aid of inert gas plasmas [76]. A plasma consists usually of a gaseous mixture of electrons, ions, excited molecules, free radicals, atoms and photons, which are embedded in the bulk of intact molecules.

The plasma is usually generated in glow discharge systems, which operate either with or without electrodes. Schematic diagrams of discharge systems are shown in Fig. 5.8. For further information the reader should consult reference [76].

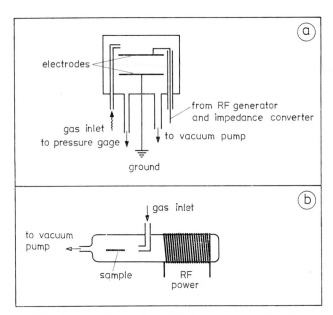

Fig. 5.8 Schematic illustration of discharge systems utilized for plasma generation. (a) System with two planar electrodes, (b) inductively coupled electrodeless system. RF = radio frequency (typical 13 MHz)

Table 5.4 Plasma treatment of polymers. Rate of weight loss in a helium plasma, 100 µm Hg at 30 W (according to *Yasuda* [77])

Polymer	Rate of Weight Loss $(10^{-3}$ mg/cm² min)
polyoxymethylene	17.0
poly(acrylic acid)	16.2
poly(methacrylic acid)	15.4
polyvinylpyrrolidone	11.9
polyvinylalcohol	9.4
polyethyleneterephthalate	1.7
polyethylene	1.2
polyamide-6	1.1
polypropylene	0.8

For the treatment of polymer surfaces, plasmas of helium, argon, N_2, O_2, and H_2 have been used. Most polymers lose weight upon exposure to a plasma [77], as can be seen from Table 5.4. Therefore, plasma etching is used to examine morphological features in surfaces of polymer specimen [81, 82] and to remove polymer coatings from underlying layers. With N_2 and O_2 plasmas, nitrogen and oxygen, respectively, are incorporated in the polymer surfaces. Oxidations in surface layers of polyethylene, induced by O_2 plasma treatment, cause an increase in wettability and facilitate printing and adhesion [78—81]. Inert gas plasma can be used to crosslink surfaces of polyethylene samples, a technique designated as CASING: Crosslinking by Active Species of Inert Gases.

For additional information on the application of plasma chemistry to polymers, the reader may refer to [76, 77, 83] and the references cited therein.

References to Chapter 5

[1] C. Kunz (ed.), "Synchrotron Radiation, Techniques and Application", Springer, Berlin (1979).
[2] H.-J. Queisser (ed.), "X-ray Optics", Vol. 22 of "Topics in Applied Physics", Springer, Berlin (1977).
[3] R. E. Watson, M. L. Perlman (eds.), "Research Applications of Synchrotron Radiation", Brookhaven National Lab., Upton, N.Y. (1973).
[4] F. A. Bovey, "The Effects of Ionizing Radiation on Materials and Synthetic High Polymers", Interscience, New York (1958).
[5] A. Charlesby, "Atomic Radiation and Polymers", Pergamon Press, Oxford (1960).
[6] J. O. Turner, "Plastics in Nuclear Engineering", Reinhold, New York (1961).
[7] A. Chapiro, "Radiation Chemistry of Polymeric Systems", Interscience, New York (1962).
[8] R. O. Bolt and J. G. Carrol, "Radiation Effects on Organic Materials", Academic Press, New York (1963).
[9] A. R. Shultz, in E. M. Fettes (ed.), "Chemical Reactions of Polymers", p. 645, Interscience, New York (1964).
[10] R. S. Alger, "Radiation Effects in Polymers", in D. Fox, M. M. Labes and A. Weissberger (ed.), "Physics and Chemistry of the Organic Solid State", Vol. II, Chapter 9, Interscience, New York (1965).
[11] R. Salovey, in S. Goldfein (ed.), "Breakdown of Plastics", Dekker, New York (1969).
[12] M. Dole (ed.), "The Radiation Chemistry of Macromolecules", Academic Press, New York (1972).
[13] J. E. Wilson, "Radiation Chemistry of Monomers, Polymers and Plastics", Dekker, New York (1974).
[14] F. A. Makhlis, "Radiation Physics and Chemistry of Polymers", Wiley, New York (1975) (Translation from Russian).
[15] W. Schnabel, "Degradation by High Energy Radiation" in: H. H. G. Jellinek, "Aspects of Degradation and Stabilization of Polymers", Elsevier, Amsterdam (1978).
[16] A. Swallow, "Radiation Chemistry", Longman, London (1973).
[17] J. H. O'Donnel and D. F. Sangster, "Principles of Radiation Chemistry", Arnold, London (1970).
[18] P. Ausloos (ed.), "Fundamental Processes in Radiation Chemistry", Interscience, New York (1968).
[19] L. G. Christophorou, "Atomic and Molecular Radiation Physics", Wiley-Interscience, London (1971).
[20] E. J. Henley and E. R. Johnson, "The Chemistry and Physics of High Energy Reactions", University Press, Washington, D.C. (1969).
[21] A. Henglein, W. Schnabel and J. Wendenburg, "Einführung in die Strahlenchemie", Verlag Chemie, Weinheim (1969).
[22] C. H. Bamford and C. F. H. Tipper (eds.), "Comprehensive Chemical Kinetics", Vol. 3, Elsevier, Amsterdam (1969).

[23] *I. V. Vereshinsky* and *A. K. Pikaev*, Introduction to Radiation Chemistry, Israel Progr. Scient. Translat. Jerusalem (1964).
[24] *H. A. Bethe* and *J. Ashkin*, in *E. Segré* (ed.), Experimental Nuclear Physics, Wiley, New York (1958), pp. 166—357.
[25] *W. Schnabel*, "Pulse Radiolysis of Polymers", Hoshasen Kagaku 26, 23 (1978).
[26] *D. T. Türmer*, J. Polym. Sci., Macromol. Rev. 5, 229 (1971).
[27] *H. Fischer*, "Magnetic Properties of Free Radicals", in *K. H. Hellwege* (ed.), Landolt-Börnstein, New Series II/1, Springer, Berlin-Heidelberg-New York (1965).
[28] *S. Ya. Pshezhetskii, A. G. Kotov, K. K. Milinchuky, V. A. Roginskii* and *V. I. Tupikov*, "EPR of Free Radicals in Radiation Chemistry", Wiley, New York (1974).
[29] *D. Campbell*, J. Polym. Sci., Macromol. Rev. 4, 91 (1970).
[30] *B. Rånby* and *J. F. Rabek*, "ESR Spectroscopy in Polymer Research", Springer, Berlin (1977).
[31] *S. K. Ho* and *S. Siegel*, J. Chem. Phys. 50, 1142 (1969).
[32] *C. David*, "High Energy Degradation of Polymers" in *C. H. Bamford* and *C. H. F. Tipper*, (eds.) "Comprehensive Chemical Kinetics", Elsevier, Amsterdam (1975).
[33] *C. David, A. Proumen-Demiddeleer* and *G. Geuskens*, Rad. Phys. Chem. 11, 63 (1978).
[34] *K. Yahagi*, "Gamma-Ray Induced Conductivity in Polymer Insulating Materials", Memoirs of the School of Science and Engineering, Waseda University 34, 71 (1970).
[35] (a) *T. Nishitani, K. Yoshino* and *Y. Inuishi*, Jap. J. Appl. Phys. 15, 931 (1976);
     (b) *K. Hayashi, K. Yoshino* and *Y. Inuishi*, Jap. J. Appl. Phys. 14, 39 (1975).
[36] *C. David, D. Fuld* and *G. Geuskens*, Makromol. Chem. 139, 269 (1970); 160, 135 and 347 (1972).
[37] *A. Chapiro*, J. Chim. Phys., Physicochim. Biol. 52, 246 (1955).
[38] *P. Alexander* and *D. Toms*, J. Polym. Sci. 22, 343 (1956).
[39] *L. A. Wall* and *M. Magat*, J. Chim. Phys. Physicochim. Biol. 50, 308 (1953).
[40] *T. N. Bowmer, L. K. Cowen, J. H. O'Donnell* and *D. J. Winzor*, J. Appl. Polym. Sci. 24, 425 (1979).
[41] *A. Chapiro*, J. Chim. Phys., Pysicochim. Biol. 53, 895 (1956).
[42] *C. Wippler*, J. Polym. Sci. **29**, 595 (1958).
[43] *L. Wall* and *R. E. Florin*, J. Appl. Polym. Sci. 2, 251 (1959).
[44] *C. H. Atchinson*, J. Polym. Sci. 35, 557 (1959).
[45] *A. Charlesby* and *W. Moore*, Int. J. Appl. Radiat. Isotop. 15, 703 (1964).
[46] *J. R. Brown* and *J. H. O'Donnel*, Macromolecules 5, 109 (1972).
[47] *K. Wündrich*, Eur. Polym. J. 10, 341 (1974).
[48] *J. R. Brown* and *J. H. O'Donnell*, J. Appl. Polym. Sci. 23, 2763 (1979).
[49] *C. E. Scroog, A. L. Endrey, S. V. Abramo, C. E. Berr, W. M. Edwards* and *K. L. Olivier*, J. Polym. Sci. A 3, 1373 (1965).
[50] *W. Schnabel*, "Application of the Light Scattering Detection Method to Problems of Polymer Degradation" in *N. Grassie* (ed.), "Developments in Polymer Degradation", Appl. Science Publ., London (1979).
[51] *D. Lindenau, S. W. Beavan, G. Beck* and *W. Schnabel*, Eur. Polym. J. 13, 819 (1977).
[52] *U. Hagen*, Biochem. Biophys. Acta 134, 45 (1967).
[53] *D. Lindenau, U. Hagen* and *W. Schnabel*, Rad. Environm. Biophys. 13, 287 (1976).
[54] *T. Coquerelle, L. Bohne* and *U. Hagen*, Z. Naturforsch b 24, 885 (1969); Intern. J. Rad. Biol. 17, 205 (1970) and with *B. Kessler*, Intern. J. Rad. Biol. 24, 397 (1973).
[55] *A. J. Bertinchamps* (ed.), "Effects of Ionizing Radiation on DNA", Springer, Berlin (1978).
[56] (a) *L. S. Myers*, "Radiation Chemistry of Nucleic Acids, Proteins, and Polysaccharides" in ref. [12], Vol. II, Chapter 17;
     (b) *A. J. Swallow*, "Radiation Chemistry", Longman, London (1973);
     (c) *H. Dertinger* and *H. Jung*, "Molecular Radiation Biology, Springer, Berlin (1970).
     (d) *K. I. Altman, G. B. Gerber* and *S. Okada*, "Radiation Biochemistry", Academic Press, New York (1970).

[57] *M. J. Bowden* and *L. F. Thompson*, Solid State Technol. 22, 72 (1979).
[58] *J. H. Lai, L. Shepherd, R. Ulmer* and *C. Gier*, Proc. 4th Photopolym. Conf. Soc. Plast. Eng. Ellenville, N.Y. (1976).
[59] *E. Gipstein, W. Moreau,* and *O. Need,* J. Electrochem. Soc. 123, 1105 (1976).
[60] *Y. Hatano, H. Morishita,* and *S. Nonogaki,* ACS Div. Org. Coat. and Plast. Chem. Preprints 35, 258 (1975).
[61] *S. Matsuda, S. Tsuchiya, M. Houma, K. Hasegawa, G. Nagamatsu* and *T. Asamo,* Proc. 4th Photopolym. Conf. Soc. Plast. Eng., Ellenville, N. Y. (1976).
[62] D.O. 2363092 Philips, 1972; D.O. 2610301, Philips (1975).
[63] *E. D. Roberts,* ACS Div. Org. Coatings and Plast. Chem. Preprints 33, 309 (1973); 35, 281 (1975); 37, 36 (1977).
[64] *M. Kakuchi, S. Sugawara, K. Murase* and *K. Matsugama,* J. Electrochem. Soc. 124, 1648 (1977).
[65] *M. J. Bowden* and *L. F. Thompson,* J. Appl. Polym. Sci. 17, 3211 (1973); J. Electrochem. Soc. 120, 1722 (1973); 121, 1620 (1974).
[66] *R. J. Himics, N. Desai, M. Kaplan* and *E. S. Poliniak,* ACS Div. Coatings and Plast. Chem. Preprints 35, 266 (1975).
[67] *L. F. Thompson, L. D. Yau, E. M. Doerris* and *L. E. Stillwagon,* paper presented at 8th Int. Conf. Electron and Ion Beam Sci. and Technol., Electrochem. Soc. Meet. Seattle (1978).
[68] *L. F. Thompson, L. E. Stillwagon* and *E. M. Doerris,* J. Voc. Sci. Technol. 15, 938 (1978).
[69] *T. Hirai, Y. Hatano* and *S. Nonogaki,* J. Electrochem. Soc. 118, 669 (1971).
[70] *E. D. Feit, R. D. Heidenreich* and *L. F. Thompson,* ACS Div. Org. Coatings and Plast. Chem. Preprints 33, 3383 (1973).
[71] (a) *G. Schmahl, D. Rudolph* and *B. Niemann,* "*X*-Ray Microscopy", in *H. Stuhrmann* (ed.), "Uses of Synchrotron Radiation in Biology", Academic Press, London (in press).
     (b) *Y. Farge* and *P. D. Duke* (eds.), "European Synchrotron Radiation Facility", Suppl. I "The Scientific Case", European Science Foundation, Strasbourg (1979).
     (c) *H. B. Stuhrmann,* "The Use of *X*-ray Synchrotron Radiation for Structural Research in Biology", Quarterly Rev. Biophys. 11, 71 (1978).
[72] *E. Spiller, R. Feder, J. Topalian, A. N. Broers, W. Gudat, B. J. Panessa, Z. A. Zadunaisky* and *J. Sadat,* Science 197, 259 (1977); J. Appl. Phys. 47, 5450 (1976).
[73] *E. Spiller* and *R. Feder,* "*X*-Ray Lithography", Chapter 3 in ref. [2].
[74] *V. T. Kagiya* and *M. Hagiwara,* "Radiation Recycling Process of Waste Polytetrafluoro-ethylene Scrap", IAEA-SM-194/704.
[75] *R. J. Ceresa* (ed.), "Block and Graft Copolymerization", Wiley, London (1973).
[76] *M. Shen* (ed.), "Plasma Chemistry of Polymers", Dekker, New York (1976).
[77] *H. Yasuda,* "Plasma for Modification of Polymers" in [76], p. 15.
[78] *D. H. Reneker* and *L. H. Bolz,* "Effect of Atomic Oxygen on the Surface Morphology of Polyethylene" in [76], p. 231.
[79] *R. H. Hansen, J. V. Pascale, T. De Benedicts* and *P. M. Rentzepis,* J. Polym. Sci. A 3, 2205 (1965).
[80] *J. J. Dietl,* Kunststoffe 59, 792 (1969).
[81] *A. Aldrian, E. Jakopic, O. Reiter* and *R. Ziegelbecker,* Radex Rundsch. 2, 510 (1967).
[82] *E. W. Fischer,* Kolloid Z. Z. Polym. 226, 30 (1968).
[83] *R. S. Thomas,* in *J. R. Hollahan* and *A. T. Bell* (eds.), "Techniques and Applications of Plasma Chemistry", Wiley-Interscience, New York (1974).

6   Biodegradation

6.1   Introduction

Biodegradation of polymers is familiar to everybody as far as natural polymers are concerned. We all are very well aware of the fact that living organisms cannot only synthesize biopolymers such as proteins, nucleic acids, polysaccharides (including cellulose) but are also capable of degrading them. It is quite obvious that all natural polymeric products, even wood and ivory, will eventually be decomposed into small molecules upon the death of the producing organism. Commonly, the nature of the decomposition products allows their utilization by other organisms for energy production or synthesis of new compounds (including biopolymers). Only in exceptional cases are natural decomposition processes forced to enter different routes ending up as products such as coal, peat or crude oil, which can be considered as incompletely degraded decomposition products of biopolymers and related materials.

The general mechanism of degradation of polymers into the small molecules employed by nature is a chemical one. Living organisms are capable of producing enzymes which can attack biopolymers. The attack is usually specific with respect to both the enzyme/biopolymer couple and the site of attack at the polymer. Thus, the formation of specific decomposition products is guaranteed.

With the advent of synthetic polymers the question arose: how would nature behave towards man-made polymeric materials? The answer is that the majority of synthetic polymers is rather inert towards biological enzymatic attack although, in principle, being biodegradable. While initially this fact was considered as favorable, new problems developed subsequently as the production of synthetic polymers increased tremendously and the amount of plastic waste in rubbish dumps grew simultaneously. The plastic bottle or foil becoming the 20th century's "leitfossil" (characteristic fossil) for future archaeologists is a fascinating idea which is correlated, however, with quite serious ecological problems for us and our contemporaries. Therefore, extensive research programs are under way aiming at (a) a better understanding of biological degradation processes and (b) the synthesis of new biodegradable polymers. The various modes of biological degradation concerning both biopolymers and synthetic polymers will be dealt with in the following section. Aspects concerning the biodegradability of synthetic polymers have also been discussed in several books and articles [1—3].

6.2   Modes of Biological Degradation

Generally, natural and synthetic polymers can be attacked by living organisms either chemically or mechanically.

The chemical mode pertains to the decomposition of polymers in the digestive tracts of highly organized living species, humans, for example, or to the attack of micro-organisms. Commonly, enzymes are involved in the chemical mode of polymer degradation.

During long-lasting evolutionary processes, highly organized living species were formed that became adapted to the decomposition of certain natural polymers, for instance, proteins. The decomposition processes employed are highly specific and are not operative towards substrates of different chemical nature.

Microorganisms, e.g., bacteria and fungi, behave differently. Although they are commonly also quite specific with respect to the degradation of the substrate, many of them are, however, capable of adapting to the substrate. It is known that although a great number of microorganisms can produce a variety of enzymes, microorganisms usually specialize on the attack of only a single substrate and are, therefore, producing only one or a few enzymes. If the substrate is changed, the microorganisms start after a few weeks or months the production of new enzymes capable of attacking the new substrate.

The capability of microorganisms to adapt to new substrates is, of course, of great importance to the problem of the biodegradability of synthetic polymers. It is now generally accepted, that a great number of microorganisms is capable of attacking synthetic polymers. Typical examples will be given in section 4 and the general importance of this fact will be discussed in section 6 of this chapter. At this point it must be emphasized only, that, generally, synthetic polymers are biodegradable. The reason we are not usually confronted with problems caused by microbial attack of synthetic polymers is simply because this class of materials is in principle brand new to nature. Thus, it can be anticipated that the situation will become different after the realm of microorganisms has adapted more generally to man-made polymers.

The mechanical mode of biodegradation of polymers pertains to the attack by certain mammals (e.g. rodents) and insects. Regarding materials composed of natural polymers, wood and wool, for instance, the attack by animals is a serious problem. In a number of cases the reasons for attacking the polymer include also the nutritional needs of the attacking mammal or insect (e.g. death-watch beetles and termites digest wood or moths eat wool). Synthetic polymers (e.g. polyethylene or polystyrene), on the other hand, are not attacked for reasons of nutrition. The attacking animal bites or chews articles made of synthetic polymers because the physical properties of the polymeric material are compatible with the natural needs of the animal. A typical example is the need of rodents for biting, which can cause serious problems, e.g. for plastic insulations of electrical cables placed in the ground.

6.3   Enzymatic Degradation

6.3.1   Biopolymers

Enzymes are proteins of more or less complicated chemical structure and quite different molecular weights ranging from 10^3 to 10^6. They possess hydrophilic groups ($COOH$, OH, NH_2) and are commonly soluble in aqueous systems. Precipitation is usually caused by high concentrations of univalent salts, and low concentrations of multivalent metal ions. The catalytic activity of enzymes is normally related to a special molecular conformation. Conformational changes induced by variation of pH, addition of organic solvents or temperature increase lead to "denaturation", i.e. to a loss of catalytic activity.

Some enzymes require for their catalytic activity the simultaneous presence of a coenzyme (primer) and others in addition an activating inorganic ion. In certain cases it has been found that these partners are forming association complexes. The coenzyme commonly functions as donor or acceptor of a specific group.

Frequently, enzymes are designated according to their mode of action. Hydrolases, for instance, are enzymes catalyzing the hydrolysis of ester-, ether- or amide (peptide)-linkages. Proteolytic enzymes (hydrolyzing proteins) are called proteases and enzymes hydrolyzing polysaccharides (carbohydrates) are called carbohydrases. A compilation of several typical enzymes, together with their substrate polymers is presented in Table 6.1.

Table 6.1   Enzymes capable of rupturing main chains in natural polymers

	Enzyme	Polymer	Occurrence
carbo-hydrases	amylase	amylose	bacteria, malt, pancreas
	phosphorylase	amylose, amylopectin	bacteria, yeast, animals, plants
	cellulase	cellulose	bacteria, fungi
	lysozyme	polysaccharides in cell walls	bodily secretions, whites of eggs
proteases	pepsin	proteins	gastric mucosa
	trypsin	proteins	pancreas
	carboxy-peptidase	proteins	bacteria, pancreas
esterases	ribonucleases	ribonucleic acid (RNA)	bacteria, plants, spleen, pancreas
	deoxyribo-nucleases	deoxyribonucleic acid (DNA)	bacteria, pancreas
	phosphodi-esterases	nucleic acids (DNA, RNA)	snake venom, intestinal mucosa, leukemia cells, spleen

The remainder of this section is devoted to the description of the function of several typical enzymes during the hydrolysis of polysaccharides, proteins and nucleic acids. The situation encountered with synthetic polymers will also be discussed briefly. For detailed and more comprehensive information on enzymatic processes the reader is referred to several books and articles [4—9].

6.3.1.1 Polysaccharides

α- and β-amylases are well-known enzymes active towards starch and glycogen. They hydrolyze amylose and amylopectin (constituents of starch and glycogen, the two storage carbohydrates of plants and animals). Amylose is a linear α-(1−4) glucoside:

Amylopectin is an α-(1−4) glucoside with branches commencing at the 6-positions

Amylases operate in two modes as either endo-(α-amylases) or exo-amylases (β-amylases). α-amylases hydrolyze only (1−4) linkages and attack the amylose chains at random points. The end products are α-maltose (a disaccharide) and glucose:

α-maltose

α-glucose
(α-D-glucopyranose)

α-maltose is not attacked by α-amylases. β-amylases also hydrolyze only (1−4) linkages, but attack amylose specifically only at the non-reducing end of the chain removing successively maltose molecules. This is illustrated in Fig. 6.1.

Since neither α- nor β-amylases can attack (1−6) linkages, the end products of the hydrolysis of amylopectins contain branched dextrins with intact (1−6) linkages.

The polysaccharide phosphorylases are also noteworthy. They catalyze the phosphorolytic cleavage of α-D-glucose-1-phosphate from the non-reducing end of the amylose chain [10]

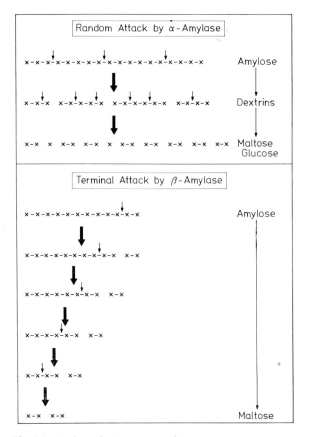

$$+ \text{ H}_3\text{PO}_4$$

$$(6.1)$$

Phosphorylase

(Cori ester)

Fig. 6.1   Action of α-(endo) and β-(exo) amylase on amylose. —X—X— represents glucose repeating units connected by (1—4) linkages

As indicated by the reversibility of reaction (6.1), phosphorylases are also capable of catalyzing the synthesis of amylose from α-D-glucose-1-phosphate.

Cellulases are enzymes catalyzing the hydrolysis of cellulose. They play an important role in the natural decomposition of plant residues and are produced by a great number of cellulolytic microorganisms including aerobic saprophytes, anaerobic rumen bacteria and anaerobic thermophilic spore formers.

Cellulose is a linear β-(1−4) glucoside:

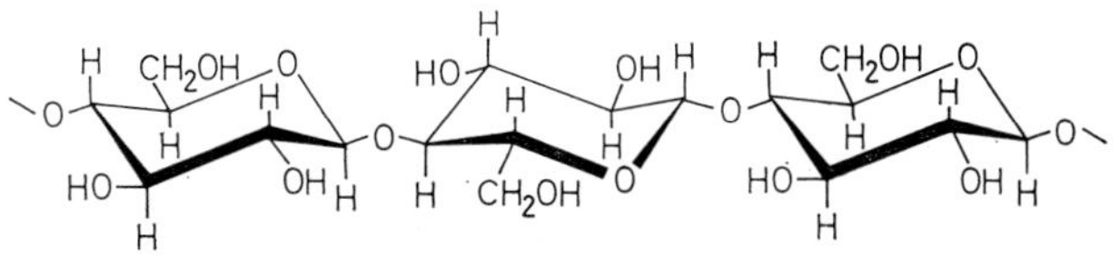

The cellulases operate in two modes analogous to the amylases [4] as either endo- or exocellulases. A cellulase obtained from the mold Myrothecium verrucaria attacks cellulose at random (endocellulase) and produces cellobiose and glucose as end products:

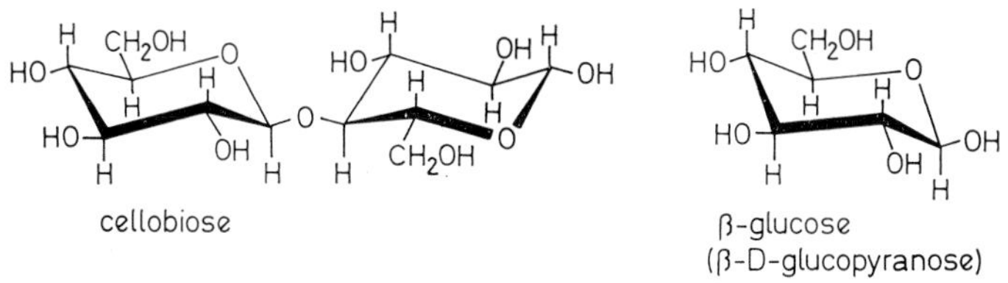

cellobiose

β-glucose
(β-D-glucopyranose)

Other cellulases (exocellulases) were found to attack cellulose in a way analogous to the mode of action of β-amylase. They begin to digest cellulose at the non-reducing end of the macromolecule and attack the second or the third glucoside linkage from the end thus producing cellobiose or cellotriose, respectively.

Another interesting example refers to lytic enzymes (e.g. lysozyme) produced by phages with the purpose of destroying cell walls of bacteria. Such cellwalls consists essentially of peptidoglycans composed of glycan strands linked together by peptide bridges. The glycan is a polysaccharide: an alternating copolymer of N-acetylglucosamin and N-acetyl-muramic acid.

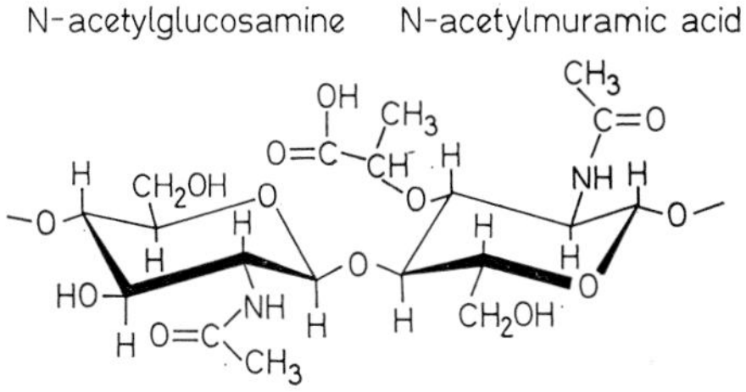

N-acetylglucosamine   N-acetylmuramic acid

The mode of action of various lytic enzymes is depicted in Scheme 6.1. Endoacetyl-muramidases (e.g. produced by phage T4 or T2) split the bond (b) of N-acetylmuramyl-N-acetylglucosamine whereas endoacetylglucosaminidases (e.g. streptococcal muralysin)-

rupture the N-acetylglucosaminyl-N-acetylmuramic acid bond (a). Other enzymes (peptidases) are capable of splitting peptide bonds (c), as indicated in Scheme 6.1.

Scheme 6.1   Mode of action of lytic enzymes during the lysis of peptidoglycan (mucopeptide), the main constituent of bacteria cell walls (according to *A. Tsugita* [40])

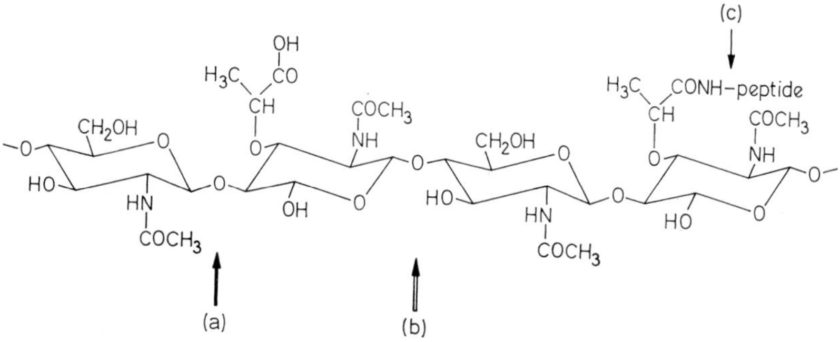

To conclude this section we should mention some technical applications of enzymatic degradation processes. For example the enzymatic degradation of polysaccharides is utilized in the textile industry. Starch, used as a sizing agent for weaving of yarn, is subsequently removed by amylases (from pancreas or bacteria) available as commercial products from various companies [27, 28].

Furthermore, various biotechnical processes have become important recently with regard to the utilization of wood and straw as sources of polysaccharides [11—15]. Apart from cellulose, these natural products contain other valuable polysaccharides (xylans and mannans) which are attacked by specifically acting enzymes (xylanases, mannanases) in a similar mode as cellulose is attacked by cellulases. In order to carry out these processes on technical scales difficulties concerning delignification have to be overcome. Lignin is an important constituent of wood and has to be separated from the other components. A rather gentle method of decomposition of wood refers to steam treatment permitting, in a consecutive step, the extraction of polysaccharide fractions by washing with water or aqueous alkaline solution [12]. The extracted polysaccharides can then be subjected to enzymatic attack leading to low molecular weight sugars. The possibility of enzymatic degradation of otherwise undigestible polysaccharides opens up new routes to sources of nutrients for animal feeding and, therefore, to food. Contrary to chemical processes, biotechnical, i.e. enzymatic, processes operate under gentle conditions (no harmful chemicals are used) at moderate temperatures and normal pressure, thus meeting important environmental demands.

6.3.1.2   Proteins

Proteases act during the digestion of protein containing food in the body of higher developed living organisms. Moreover, they play an important role during putrefaction of the organisms after death. They are produced by certain organs in the body of living organism or by microorganisms. Proteins are copolymers *) consisting of up to 20 different L-α-amino acids of the structure $H_2N-CHR-COOH$ connected by peptide bonds:

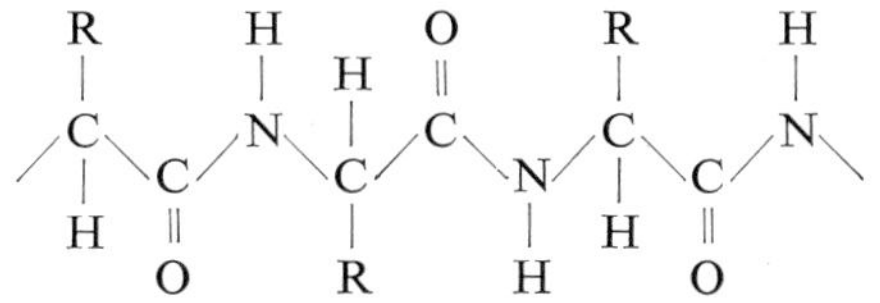

with R: $-H$, (Glycine), $-CH_3$ (Alanine), $-(CH_2)_4-NH_2$ (Lysine), $-CH_2-SH$ (Cystein) etc.

Copolymers of MW lower than ca. 10^4 are called polypeptides, those of higher MW are denoted as proteins.

Pepsin and chymotrypsin attack peptide linkages at random, whereas trypsin attacks the carbonyl groups of arginine and lysine only. Carboxypeptidase acts specifically at free carboxyl groups at chain ends and splits off terminal amino acids. The enzymatic hydrolysis of proteins serves as a very useful aid in determining sequences of amino acids in proteins.

In general, proteases are more active against denatured proteins than against native proteins.

Numerous microbial processes are known which cause a further decomposition of the amino acids formed during hydrolysis of proteins [4]. The fermentation of alanine yields propionic acid, acetic acid, CO_2 and NH_3. Arginine is cleaved into ornithine, CO_2 and NH_3, glutamic acid into acetic acid, CO_2, NH_3 and H_2, and glycine into acetic acid, CO_2 and NH_3.

6.3.1.3   Nucleic Acids

Ribonucleic acids (RNA) and deoxyribonucleic acids (DNA) are polynucleotides, i. e. polyesters consisting of nucleotides of the structures:

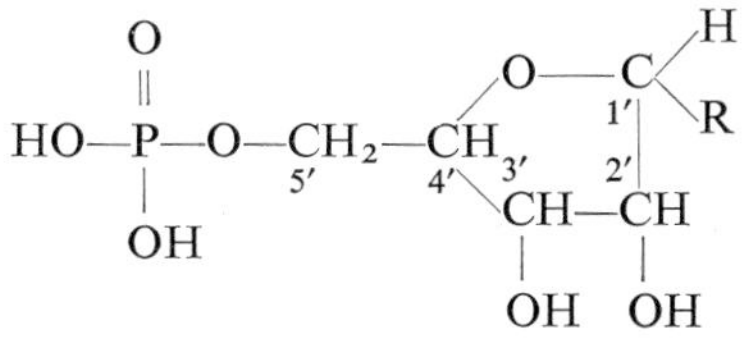

ribonucleotide 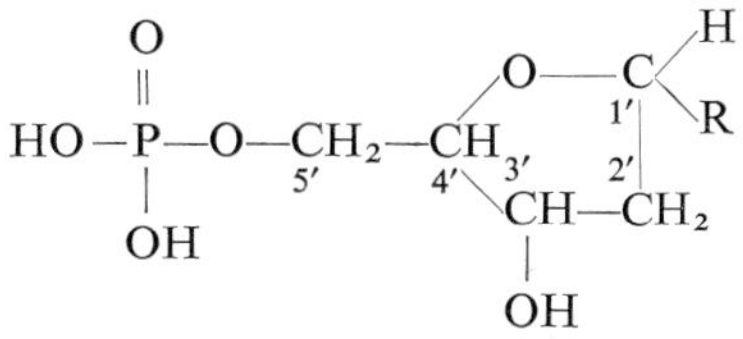   deoxyribonucleotide

with R generally representing adenyl, cytosyl, guanyl, thymyl. The nucleotides are connected via 3′ and 5′ ester linkages. A typical structure of DNA is given below:

---

*) copolyamides

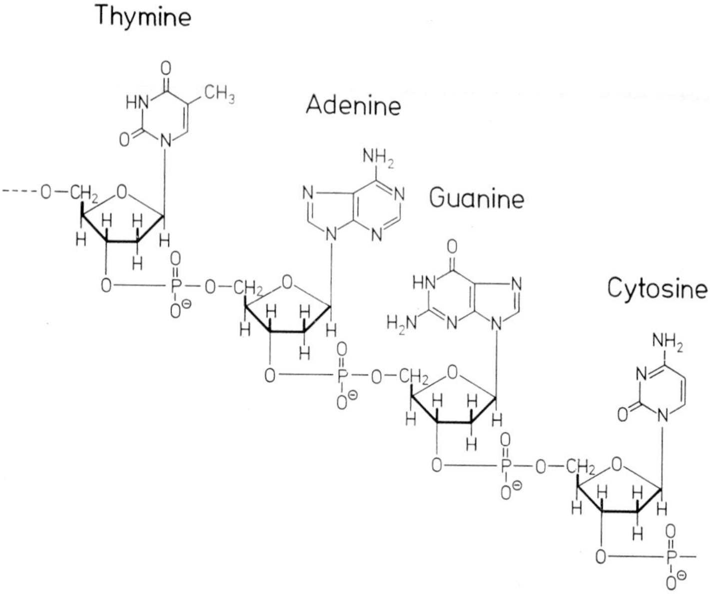

Thymine

Adenine

Guanine

Cytosine

In the native state two DNA chains are combined via hydrogen bonds to form a double stranded helix. DNA chains have a MW of up to several hundred millions; the MW of RNA chains is about 2×10^4 to 10^5. Esterases, attacking the internucleotide linkages in polynucleotides (nucleic acids), fall into three main groups: ribonucleases (RNases), deoxyribonucleases (DNases) and phosphodiesterases. These enzymes catalyze the cleavage of polynucleotides into nucleotides. The latter can be decomposed further by the catalysis of nucleotidases into nucleosides and phosphoric acid. This is shown in Scheme 6.2.

Scheme 6.2   Enzymatic hydrolysis of polynucleotides

---

polynucleotide

   | nuclease
   ↓

polynucleotide fragments

   | nuclease
   ↓

nucleotides $\xrightarrow{\text{nucleotidase}}$ H_3PO_4 + nucleoside

---

Whereas RNases and DNases only attack the phosphate ester bonds in the corresponding nucleic acid (i.e. RNA and DNA, respectively), phosphodiesterases catalyze the hydrolysis of both RNA and DNA chains. Commonly, phosphodiesterases act as exonucleases, i.e. they attack the polynucleotide chains at one end and split off mononucleotides. RNases and DNases are endonucleases, i.e. they attack exclusively phosphate ester linkages in the interior of the polynucleotide chains.

Specificity correlated to the chemical structure of the bases has also been detected. For instance, RNase was reported to catalyze the cleavage of ester linkages only at nucleotides containing pyrimidin bases (uracil and cytosine) and not at nucleotides containing purine bases (adenine and guanine) [5].

For further details the reader is referred to the "enzyme literature", e.g. references [5, 8, 9].

The mode of action of nucleases on polynucleotides is demonstrated by a typical example in Scheme 6.3.

Scheme 6.3   Mode of action of ribonuclease

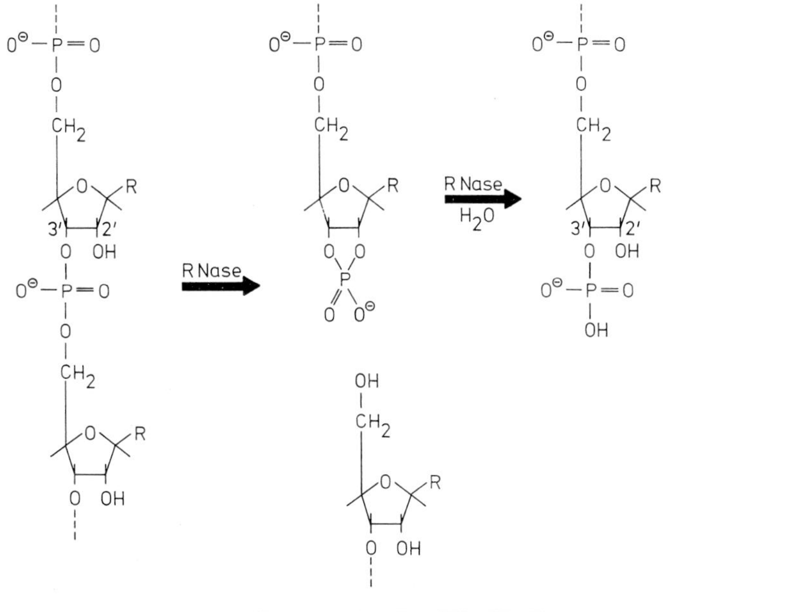

In this case ribonuclease first produces a 2′,3′-phosphoric diester ring which then opens at a later stage.

6.3.2   Synthetic Polymers

Most of the present knowledge on microbial decomposition of synthetic polymers is based on studies with microorganisms (see Section 6.4). Only occasionally have systematic studies been carried out with the aim of elucidating the interaction of enzymes with synthetic polymers. Noteworthy is work related to the synthesis of biodegradable polymers. This will be discussed in Section 6.5.

Of general importance is the work of Baptist et al. [16]. They investigated the oxidation of a linear low molecular weight hydrocarbon, n-octane, with a cell-free enzyme preparation from Pseudomonas oleovorans. Octanol, octaldehyde and octanoate were

11*

formed. Evidence for the action of mixed-function oxidases and of alcohol dehydro-genase was obtained. That work, and other studies with somewhat longer hydrocarbon chains [17], demonstrated that enzymatic oxidation of hydrocarbons occurs at the chain end only, a finding which has been corroborated by investigations on the microbial decomposition of linear polymeric hydrocarbons of moderate chain lengths (see Section 6.4).

It appears that, owing to their chemical composition and their configuration, many synthetic polymers are not sensitive to lateral attacks (at sites far away from the chain ends) by enzymes commonly generated by microorganisms. This explains the high resistance of many polymers against microbial attack.

6.4   Microbial Degradation of Synthetic Polymers

Microorganisms play an eminent role in the decomposition of organic material of all kinds including biopolymers. There is a great body of microorganisms, such as fungi, bacteria and actinomycetes, which are distributed ubiquitously around the earth. Under appropriate conditions the growth of microorganisms will occur simultaneously with the decomposition of any biological species after the latter has perished. In many cases the death of living biological organisms is caused by the attack of microorganisms, e.g. pathogenic bacteria. During rotting of organic material, e.g. in compost heaps, microorganisms are also operative. The growth of microorganisms depends on pH, temperature, availability of mineral nutrients, O_2 concentration and humidity (the presence of water being a prerequisite).

Fungi (e.g. molds) commonly require O_2 and a pH of $4.5-5.0$ to proliferate. They grow over a wide temperature range up to 45 °C. Optimum growing rates occur in most cases at temperatures between 30° and 37 °C.

Actinomycetes generally grow aerobically in the pH range between 5 to 7. They are usually mesophilic with regard to temperature, i.e., they will grow over a wide range of temperature.

Bacteria are either aerobic or anaerobic and grow in a pH range between 5 and 7. They are also mesophilic with regard to temperature.

Some microorganisms are thermophilic, i.e. they proliferate at rather high temperatures ($40°-70$ °C) with optimum growing rates at $50°-55$ °C, or even higher at $50°-70$ °C. Temperatures appreciably higher than room temperature frequently prevail in compost heaps. Furthermore, thermophilic microorganisms have attained importance in the decomposition of plastic material in waste-bioreactors [38 (a) and (c)] (see Section 6.6). Table 6.2 contains a list of microorganisms which proliferate at temperatures above 60 °C.

Commonly, the microbial degradability of synthetic polymers is studied by growth tests on solid agar media.*) Test fungi and/or test bacteria are inoculated together with the polymeric material (the latter in the form of films, granules, plaques or powder).

---

*) in USA: Standard tests ASTM-D-1924-63 and ASTM-D-2676T, in other countries: DIN 53933, NFX 41-514, EMPA VII/11, Vitno Bio A_1, Iso A, Iso B, CSN 038180 A

Table 6.2   A selection of thermophilic, aerobic microorganisms existing in soil and being capable of proliferating at temperatures above 60 °C (adapted from ref. [38 (c)]

	species	optimum growth temperearur (°C)
Bacillus	calidolactis	60—65
	pepo	60
	therminalis	60—65
	thermodiastaticus	65
	tostus	60—70
Actinomyces	nondiastaticus	65
(Streptomyces)	spinosporus	60—65
	thermodiastaticus	65
	thermofuscus	60
	thermophilus	60
Fungi:		
Thermoascus	aurantiacus	60
Thermoidium	sulfureum	60

The agar media contain all nutrients necessary for microbial growth except a carbon source. Typical microorganisms employed for these tests are listed in Table 6.3. The tests are run over a definite time (usually 3 weeks). Growth rates are classified according to the fraction of the gel-surface covered with colonies, i.e. 0: no growth; 1: 10% covered; 2: 10—30% covered; 3: 30—60% covered; 4: 60—100% covered. For more quantitative tests, polymer layers are deposited on the bottom of Petri dishes and covered with nutrient agar medium. After microbial treatment, weight changes and other physical or chemical alterations are checked for. In the so-called clear zone test polymer powder is suspended in a nutrient agar medium. The hazy gel becomes clear in places surrounding colonies indicating that the polymer has become soluble due to enzymatic attack.

Table 6.3   Typical microorganisms employed for biodegradability tests of synthetic polymers

Fungi	Aspergillus niger, Aspergillus flavus, Chaetomicum globosum, Penicillium funiculosum, Pullularia pullulans
Bacteria	Pseudomonas aeruginosa, Bacillus cereus, Coryneformes bacterium, Bacillus sp.
Actinomycetes	Streptomycetaceae

Soil burial tests have been employed rather frequently in order to check for polymer weight losses or MW changes caused by microbial attack. Recently, ^{14}C tracer studies have been carried out in the investigation of biodegradation processes causing the evolution of CO_2. In these cases ^{14}C-labelled polymers were used and the microbial attack could be evidenced even in the presence of additional (unlabelled) carbon sources. The evolved CO_2 is absorbed in alkaline solution, a procedure which allows determination of the total amount of CO_2 evolved (by titration) and the fraction of CO_2 stemming from the polymer (by measuring the decay rate of the radioactivity due to $^{14}CO_2$).

This section is devoted to the microbial degradability of important synthetic polymers produced on a large scale in various countries around the globe. Biodegradable manmade polymers will be dealt with in section 5 of this chapter.

Table 6.4 lists typical degradation results obtained with standard tests [18]. It is seen that most plastics were rather resistant. Polyethylene and polyvinylchloride were found to become more resistant towards microbial attack after extraction with toluene. These findings show that low molecular weight additives (in the case of PVC the soybean oil plasticizer) are virtually degradable whereas the macromolecular matrix is scarcely or not affected. Since many commercial plastics contain additives, the findings obtained with the two samples mentioned here are of general importance. Actually, microbial growth detected on polymer samples is quite frequently due to the interaction of microorganisms with plasticizers, stabilizers etc. rather than with the macromolecules themselves. Potts et al. [19] have measured the biodegradability of many additives used in commercial plastics.

As many microorganisms are capable of producing hydrolases (enzymes catalyzing hydrolysis), it was assumed that polymers containing hydrolyzable groups in the main chains would be especially prone to microbial attack. This assumption proved to be rather useful for the development of concepts for the synthesis of biodegradable polymers (see Section 6.5).

As a matter of fact only aliphatic polyesters, polyethers, polyurethanes and polyamides exhibit a quite general sensitivity towards commonly occurring microorganisms. The incorporation of side groups, or the substitution of existing side groups by other groups usually causes inertness. This holds also for the modification of biodegradable natural polymers. Acetylation of cellulose or vulcanization of natural rubber makes these polymers quite stable against microbial attack. Moreover, biodegradability is usually strongly influenced by branching and chain length. This is so because of the specific action of enzymes with respect to configuration and chemical structure. The reader is reminded of the enzymatic decomposition of polysaccharides (see Section 6.3), where certain enzymes are incapable of attacking repeating units carrying branches, and others attack the polysaccharide chains at the terminal base units only. In the case of hydrocarbons (paraffins and polyethylene), the influence of chain length and branching on biodegradability has been rather intensely studied. The results have been reviewed recently by Potts [2]. As shown in Table 6.5, biodegradation was observed with linear paraffins at $MW \leq 450$. — Branched and high molecular weight ($MW > 450$) hydrocarbons were not affected.

Since small extents of degradation cannot be detected with the aid of growth rate tests, a more refined method was used in order to find out whether high molecular weight

Table 6.4   Microbial degradation of commercial plastics
(adapted from *Potts* et al. [18])

product	growth rating *) on agar plates
polyisobutene, poly-4-methyl-1-pentene, polymethylmethacrylate, poly(vinyl butyral), polyformaldehyde, poly(vinyl ethyl ether), cellulose acetate, bisphenol A polycarbonate, ABS terpolymer	0
poly(vinyl acetate), styrene-butadiene block copolymer poly(vinylidene chloride), poly(ethylene terephthalate), polystyrene, polypropylene	1 (less than 10% covered)
polyethylene	2 (10—30% covered)
plasticized polyvinylchloride	3 (30—60% covered)
polyurethane	4 (60—100% covered)

*) standard tests with Aspergillus niger, Aspergillus flavus, Chaetomium globosum, and Penicillium funiculosum

Table 6.5   Microbial degradation of hydrocarbons.
Influence of molecular weight and branching (after *Potts* et al. [19])

hydrocarbon	MW	branched	growth rating on agar plates *)
n-dodecane ($C_{12}H_{26}$)	170	no	4
2,6,11-trimethyldodecane ($C_{15}H_{32}$)	212	yes	0
n-hexadecane ($C_{16}H_{34}$)	226	no	4
2,6,11,15-tetramethylhexadecane ($C_{20}H_{42}$)	283	yes	0
n-tetracosane ($C_{24}H_{50}$)	339	no	4
squalane ($C_{30}H_{62}$)	423	yes	0
n-dotriacontane ($C_{32}H_{66}$)	451	no	4
n-tetracontane ($C_{40}H_{82}$)	563	no	0
n-tetratetracontane ($C_{44}H_{90}$)	619	no	0

*) for denotation see Table 6. 4.

polyethylene is susceptible to microbial attack. The evolution of $^{14}CO_2$ from ^{14}C-labelled polyethylene buried in soil was measured [21—23]. This method has been described earlier in this section. It appears that there is a limited microbial conversion of ^{14}C to $^{14}CO_2$ by soil fungi, in accordance with conclusions arrived at earlier [24], that low molecular weight oligomers present in commercial polyethylene are susceptible to microbial attack, whereas high molecular chains are not affected.

On subjecting ^{14}C-labelled polyethylene to UV irradiation prior to soil burial treatment the extent of degradation leading to $^{14}CO_2$ increased, especially if the polyethylene contained an additive promoting the photochemical oxidation [21, 22].

With respect to the biodegradability of high molecular weight synthetic polymers the ability of microorganisms to adapt to new sources of nutrients is highly noteworthy. Evidence of adaptation has been obtained in several cases. With polyamide-6, for example, it was found that Pseudomonas aeruginoa started proliferation 56 days after the bacteria were brought into contact with the polymer. Innoculation of these bacteria to previously untreated polyamide resulted in immediate growth on the new substrate [20]. Another example concerns polyvinylalcohol (PVA) [25, 26]. Certain strains of Pseudomonas, Xanthomonas and other bacteria selected from oxygen-activated sludge are capable of decomposing this water-soluble polymer. It takes about 12 days until biodegradation becomes operative, after which time microbial attack starts immediately on adding new polymeric material. If the addition of PVA is interrupted for more than 6 days the bacteria start losing the capability of decomposing PVA, but readaptation is possible [27].

In the textile industry, where PVA is used as a sizing agent, the microbial decomposition of PVA has gained technical importance in waste water purification [28]. Contrary to the situation encountered with readily biodegradable sizing agents (starch and certain cellulose derivatives) the utilization of PVA frequently caused environmental problems.

In conclusion, it should be pointed out that, in principle, synthetic plastics are susceptible to microbial attack. However, apart from a few exceptional cases, there are no microorganisms prepared to readily attack synthetic plastics, although many are capable of adapting. Also, one should be careful in drawing too general conclusions, as it is not yet clear whether the ability to adapt refers really to all synthetic polymers. Appropriate growing conditions are most important for the proliferation of microorganisms resulting eventually in the total conversion of nutrients. In the case of polymers being the carbon source (this pertains to synthetic as well as to biopolymers) the conversion of the nutrients corresponds to the decomposition of the polymeric material. Due to the lack of appropriate growing conditions, e.g., microorganisms could not have become operative in the burial-places of the Pharaohs in Egypt where moisture was lacking. It is because of the hydrophobicity of many important plastics, that microbial growth is usually prevented or limited to a thin surface layer. This brings up another essential point: the specific surface, which is low for most plastics. Frequently comminution (increase of specific surface) leads to an acceleration of the microbial growth rate.

Furthermore, it appears that — contrary to biopolymers — synthetic polymers are rather generally attacked only at the chain ends. The quite high degradation rates observed for polyvinylalcohol and poly-ε-caprolactone seem to indicate that these two polymers are exceptional cases. Due to the preferential terminal susceptibility of most synthetic macromolecules, the rates of decomposition are pretty low, as the chain ends are quite often hidden in the polymeric matrix and become not, or only very slowly, accessible to attacking enzymes. It has to be pointed out, moreover, that, kinetically, the enzymatic digestion of long chain polymers is a one-step process (not a chain reaction) consisting of many consecutive elementary reactions. In each elementary reaction one or two base units are split off. Therefore, the average molecular weight and correlated physical properties of a sample decrease only very slowly as the enzymatic reaction proceeds. A much higher rate of physical property change is expected (at the same rate of conversion expressed, as the number of base units affected per unit time) if the polymer chains are attacked at random, i.e. if the enzyme involved functions according to the "endo" and not to the "exo" mode of action.

6.5 Synthesis of Biodegradable Polymers

Table 6.6 lists various applications of biodegradable plastics. It is seen that readily degradable polymers are desirable for a variety of reasons and that there are interesting applications in the areas agriculture and medicine. Use of biodegradable plastics in packaging would, of course, help to eleviate the litter problem. As a result of the diverse fields of applications, it is easy to see why interest in synthesizing biodegradable polymers developed after the bio-resistant nature of synthetic polymers had been recognized.

Table 6.6   Applications of biodegradable plastics

Area	Applications
Agriculture	containers for transplanting seedlings, trees and other plants
	seed cases (foils or strips with separately embedded seed-corns) permitting the homogeneous distribution of seed corns
	mulches
	microencapsulations for the slow release of fertilizers, insecticides, fungicides, nematocides etc.
Medicine	surgical sutures
	implants delivering medications in a controlled manner
Packaging	discardable containers and wrapping foils
Miscellaneous	foils for prevention of erosion at freshly cultivated hill sides — sandbags for temporary dikes to control floods

The question arises, of course, why biopolymers which are readily degradable cannot be used for the purposes listed in Table 6.6. The answer refers mainly to the fact that biopolymers cannot be easily fabricated. They decompose on heating before they melt and can be processed only from solution and not by injection molding, melt extrusion etc. Modification of natural polymers, on the other hand, resulting in a lowering of the melting point, frequently causes the materials to become non-biodegradable. The modification of cellulose by esterification is a typical example. Therefore, the task of synthesizing readily biodegradable polymers, which can be easily fabricated, remains. The various approaches pertaining to the synthesis of biodegradable polymers have been discussed in detail in several articles [2, 3, 18].

In developing synthetic routes it was necessary to match the capabilities of ubiquitous microorganisms as closely as possible. As has been pointed out in Section 6.4, the relevant mode of microbial, i.e. of enzymatic action, is hydrolysis. Polymers with hydrolyzable groups, therefore, are showing promise as biodegradable systems.

The study of many commercially important polymers has revealed that aromatic and branched polyesters and polyamides are bioresistant. Since aliphatic polyesters and polyamides showed signs of biodegradability, methods aimed at synthesizing appropriate aliphatic polymers were developed. Following another idea, incorporation of vulnerable sections into the main chains of polymers was attempted by synthesizing alternating copolyamides with one comonomer being an α-amino acid. Moreover, block copolymers were synthesized containing blocks of a biodegradable polymer. In Table 6.7 and 6.8 typical polymers specially synthesized for biodegradability tests are compiled. Aliphatic polyesters are quite readily degradable. The biodegradability of poly-ε-caprolactone (No. 1 in Table 6.7) even at a relatively high molecular weight (40000) is noteworthy. This polymer, produced commercially (Union Carbide Corporation), can be fabricated by conventional thermoplastic processes and is utilized for making degradable containers for transplanting trees etc. Thin foils, decomposing within a few days in moist soil, are useful for agricultural and packaging applications.

An attempt was made to overcome disadvantages concerning the low melting point and a relatively low tensile strength by converting poly-ε-caprolactone into a new copolymer containing ester, amide and urethane linkages according to the following reactions [34]:

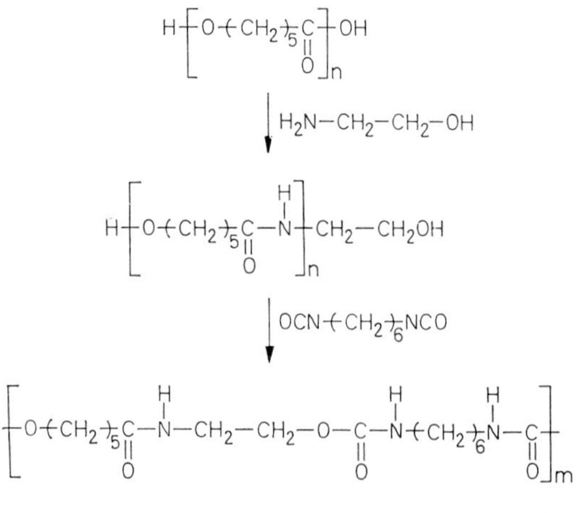

$$H-\left[O-(CH_2)_5-\underset{\underset{O}{\|}}{C}\right]_n-OH$$

$$\downarrow H_2N-CH_2-CH_2-OH$$

$$H-\left[O-(CH_2)_5-\underset{\underset{O}{\|}}{C}-\overset{H}{\underset{}{N}}\right]_n-CH_2-CH_2OH$$

$$\downarrow OCN-(CH_2)_6-NCO$$

$$\left[O-(CH_2)_5-\underset{\underset{O}{\|}}{C}-\overset{H}{\underset{}{N}}-CH_2-CH_2-O-\underset{\underset{O}{\|}}{C}-\overset{H}{\underset{}{N}}-(CH_2)_6-\overset{H}{\underset{}{N}}-\underset{\underset{O}{\|}}{C}\right]_m$$

The polymer thus obtained possesses a tensile strength about 5 times higher than that of poly-ε-caprolactone. It supports growth of various microorganisms, e.g. Aspergillus

Table 6.7 Biodegradable polyesters

No.	structural repeating unit	remarks	properties	Ref.
(1)	$-O-C(CH_2)_5-O-$ (C=O)	in soil burial tests completely degraded after one year, commercially produced	$\overline{M}_w$: 4×10^4 m.p. 63 °C, highly crystalline	[18, 19, 30]
(2)	$-O-C-CH_2-$ (C=O)	nonallergenic suture completely decomposed in the body, commercially produced		[31]
(3)	$-O-C-CH-$ (C=O, Ph)	degraded by Aspergillus niger and subtilisin		[32]
(4)	copolymers of $-O-C(CH_2)_4C-O(CH_2)_4-$ (C=O, C=O) and $-O-C(CH_2)_8C(CH_2)_4-$ (C=O, C=O)	degraded by Pullularia pullulans and other soil microorganisms	m.p. ca. 50 °C, low degree of crystallinity	[33]
(5)	$-O-C-CH-CH-C-O(CH_2)_x-$ (C=O, OH, OH, C=O) x: 2, 6, 8, 10, 12	degraded by Aspergillus niger and Aspergillus flavus		[32]

flavus and Pennicillium funiculosum and is degraded by various enzymes, e.g. rennin and urease.

Polyglycolate (No. 2 in Table 6.7), commercially available from Davis and Geck (American Cyanamid), is used as a suture of non-allergenic character which is absorbed by the body [31].

Of the remaining polymers listed in Table 6.7, poly(mandelic acid) (No. 3) is noteworthy for the fact that it is a biodegradable polymer with a pendant aromatic side group at each base unit.

No. 4 (copolymer of 1,4-butanediol with adipic and sebacid acid) and No. 5 (polyalkylene tartrates) are typical examples of various other biodegradable linear aliphatic polyesters [2, 18, 32, 33].

Table 6.8   Biodegradable Polyamides

No.	structural repeating unit	remarks	properties	Ref.
(1)	$-C(=O)-CH_2-N(H)-C(=O)-(CH_2)_5-N(H)-$	decomposed by various bacteria and fungi	hydrophilic, m.p. 268–270°C	[29]
(2)	$-C(=O)-CH(CH_2OH)-N(H)-C(=O)-(CH_2)_5-N(H)-$	readily degraded by various bacteria and fungi	water-soluble	[29]
(3)	$-C(=O)-(CH_2)_3-CH(CH_2Ph)-C(=O)-N(H)-(CH_2)_6-N(H)-$	degraded by chymotrypsin	$\overline{M}_n : 10^3-10^4$	[34]
(4)	$-C(=O)-N(H)-CH(CH_2Ph)-C(=O)-O-(CH_2)_2-O-C(=O)-CH(CH_2Ph)-N(H)-C(=O)-N(H)-(CH_2)_6-N(H)-$	degraded by chymotrypsin	$\overline{M}_n \approx 2\times10^3$	[34]
(5)	$-C(=O)-CH(Ph)-O-C(=O)-N(H)-(CH_2)_6-N(H)-$	extensive degradation by elastase or subtilisin, and by Aspergillus niger	$\overline{M}_n = 7.5\times10^3$; m.p. 110–115°C	[34]

Table 6.8 shows several biodegradable polyamides. It is interesting to note that several readily degradable copolyamides are regular alternating copolymers of an α-amino acid and ε-aminocaproic acid. The α-amino acid is glycine in No. 1 and serine in No. 2 [29]. The corresponding homopolymers polyamide-2 and polyamide-6 are rather bioresistant. Table 6.8 contains, moreover, other polyamides which are attacked by enzymes or support the growth of microorganisms: a benzylsubstituted polyamide (No. 3), a poly(ester-urea) (No. 4) and a poly(amide urethane) made from mandelic acid (No. 5).

Regarding block copolymers containing scissionable segments cellulosic derivatives are of interest [35, 36]. Scheme 6.4 presents a series of reactions leading to a copolymer prepared from cellulose and a diisocyanate.

Scheme 6.4   Synthesis of biodegradable cellulose block copolymers (after *S. Kim* et al. [36])

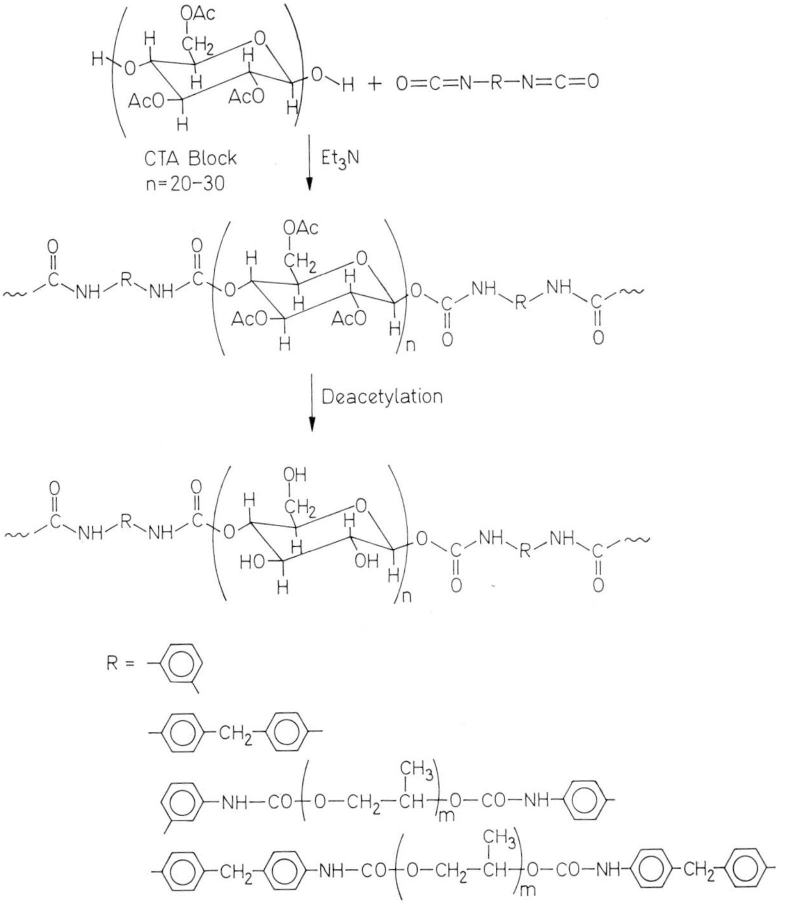

The process starts with a hydrolytic degradation of cellulose triacetate (using acetic acid) to oligomers ($\overline{DP}_n \approx 20-30$) having OH end groups. The oligomers are then reacted with diisocyanates, polyethers or polyesters, thus, a wide variety of block copolymers can be prepared. Finally, the acetyl groups capping the hydroxyl groups at the glucose ring are removed. The resulting deacetylated block copolymers are reported to undergo main-chain scission more readily than cellulose if subjected to the attack of the enzyme cellulysin at 50 °C and pH 5 [36].

A graft copolymer consisting of polymethylacrylate with side chains (MW less than 5×10^5) of starch was proposed as a material suitable for application as a biodegradable plastic mulch on vegetable fields [37]. After fabrication films are reported to possess a high mechanical strength which is rapidly lost on exposure to moisture and microbial attack. The rapid decomposition of the starch component causes extensive pitting and erosion leading finally to the comminution of the film.

6.6   Ecological Aspects

During the last two decades, people became increasingly aware of environmetal problems. Thus, municipal waste disposal also came under scrutiny. Even at the very beginning of the environmental discussions, synthetic polymeric materials, being readily detectable in all kinds of waste, attracted special attention. Actually, plastics contribute only to a limited extent to the problems of waste disposal. Worldwide, the fraction of synthetic polymers in total municipal waste is only 3 to 6%, depending on the country. Since large quantities of municipal waste are deposited, however, the absolute amount of synthetic polymers is quite large. In the Federal Republic of Germany, e.g., about 5% of the total waste is plastics, i.e. 1×10^6 tons out of 20×10^6 tons deposited annually in waste dumps. In the following paragraphs the situation concerning the fate of plastics in waste will be analyzed.

It has to be emphasized that serious litter problems are threatening the coast lines and beaches of many countries. The sea is used as a cheap and readily available rubbish dump by boats and presumably also by certain municipalities. If deposited in the sea, most plastic articles — because they do not sink — will eventually drift t o the coast.

Apart from the fact that most commercial plastics are more or less bioresistant, the conditions which prevail in most waste dumps make microbial attack rather improbable. To overcome this problem a concept based on photodegradable polymers has been developed. Accordingly, plastic packaging (bottles, foils etc.) and other plastic articles should be made from polymers possessing a limited lifetime when subjected to sunlight irradiation. The relevant chemical aspects have been outlined in Section 4.5.1. Under the influence of sunlight main-chain ruptures are induced, causing the polymer to embrittle and to eventually become comminuted. In a consecutive stage biodegradation occurs because the short chain fragments are prone to microbial attack. Although suitable photodegradable plastics have been commercially available for some time economic and other reasons have so far prevented their general utilization.

At present several modes of disposal of solid municipal waste have to be distinguished:

a) deposition in landfills (rubbish dumps),
b) composting (bioreactors),

c) incineration and
d) recycling.

Modes (a) and (b) are of interest with respect to biodegradation. By far more than 50%
of the municipal waste is deposited in landfills, only a small fraction is composted.

Whereas in former times municipal waste was frequently disposed of in "wild" rubbish
dumps (refuse heaps), modern trends are directed towards orderly disposal under aerobic
conditions. Irregular disposal under anaerobic conditions in "wild" rubbish dumps
(where the decomposition processes extend over many decades) leads to the formation
of non-oxidized decomposition products, e.g. CH_4 and H_2S. Special microorganisms,
e.g. Desulfovibrio desulfuricans, are operative. Due to the slow conversion and the
fact that systematic studies were started only about 15 years ago, our knowledge, concerning the fate of polymers in waste depositories operating under anaerobic conditions
is very limited. Digging down after several years into a 6 m deep anaerobic garbage dump
in the neighborhood of Frankfurt/Main [38 (b, c)], polymeric articles found in the upper
layers, such as polyethylene foils, polyamide stockings, cotton textiles and wood shavings,
were unaffected. Various articles located at the bottom were somewhat affected. Polyethylene bottles and foils were brittle and partly covered with Desulfovibrio colonies. Polyvinylchloride containing plasticizer had significantly deteriorated.

Uncontrolled microbial decomposition processes in anaerobic rubbish dumps might
occasionally cause pollution of the ground water and of the surrounding air.

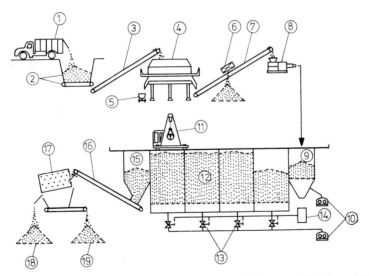

Fig. 6.2   Schematic drawing of a waste-bioreactor (from a publication of *Spohn* [39]
cited by *Wallhäusser* [38]).
(1) garbage truck, (2) storage container, (3) conveyor belt, (4) crusher, (5) coarse
garbage, (6) magnet, (7) conveyor belt, (8) homogenizer, (9) aerated transfer compartment, (10) exhaustor, (11) crane, (12) aerated reactor cells, (13) valves, (14) gas-
analyzer, (15) transfer compartment, (16) conveyor belt, (17) rotary screen, (18) coarse
refuse, (19) finished compost

In order to avoid pollution and to accelerate microbial decomposition aerobic composting procedures have been developed for industrial utilization. Fig. 6.2 shows a schematic drawing of a bioreactor system [38 (a)]. Comminuted refuse is mixed with sludge from a sewage treatment plant. Moisture and oxygen content are controlled. The composting process lasts for $10-20$ days, during which time the temperature increases to 65 to 80 °C.

It was found [38 (a)] that polystyrene, polycarbonate, polyethylene and amino- and phenolic thermosetting resins were attacked to varying extents, while other plastics remained unaffected. The observed degradation was believed to be due to the action of thermophilic microorganisms (see Table 6.2) which can prevail at these rather high temperatures. The composted material can be readily crushed and used as a soil additive, improving aeration and moisture retention.

References to Chapter 6

[1] H. H. M. Haldenwanger, "Biologische Zerstörung der makromolekularen Werkstoffe", Springer, Berlin (1970).
[2] J. E. Potts, "Biodegradation" in H. H. G. Jellinek (ed.), "Aspects of Degradation and Stabilization of Polymers", Elsevier, Amsterdam (1978).
[3] Proceedings of the 3rd Internat. Biodegradation Symposium, Kingston, Rhode Island (1975), J. M. Sharpley and A. M. Kaplan (eds.), Applied Science Publ., London (1976).
[4] J. R. Sokatch, "Bacterial Physiology and Metabolism", Academic Press, London (1969).
[5] K. J. Laidler and P. S. Bunting, "The Chemical Kinetics of Enzyme Action", Oxford University Press, London (1973).
[6] H. Morawetz, "Macromolecules in Solution", J. Wiley, New York, 2nd ed. (1975).
[7] P. Bernfeld, Adv. Enzymol. 12, 379 (1951); "Polysaccharidases" in M. Florkin and H. W. Mason (eds.), "Comparative Biochemistry", Vol. III, Academic Press, New York (1962).
[8] P. D. Boyer, H. Hardy and K. Myrbäck (eds.), "The Enzymes" (2nd edn.) Vol. 1 to 7, Academic Press, New York (1961).
[9] (a) P. D. Boyer (ed.), "The Enzymes", 3rd edition, Vol. $1-13$, Academic Press, New York $(1971-1976)$;
    (b) P. D. Boyer (ed.), "The Enzymes: Student Edition", Vol. 1 and 2, Academic Press, New York (1973).
[10] D. H. Brown and C. F. Cori, "Animal and Plant Polysaccharide Phosphorylases" in Vol. 5 of ref. [8].
[11] M. Sinner, N. Parameswaran and H. H. Dietrichs, Adv. Chem. Series 181, 303 (1979); Das Papier 32, 530 (1978).
[12] M. Sinner, N. Parameswaran, N. Yamazaki, W. Liese and H. H. Dietrichs, Appl. Polym. Symp. 28, 993 (1976).
[13] E. Gruber, Das Papier, 32, 526 (1978).
[14] B. Philipp, V. Jacopian, F. Loth, W. Hirte and G. Schulz, Das Papier 32, 538 (1978).
[15] J. Boutelje, K.-E. Eriksson and B. H. Hollmark, Svensk Papperstidning 74, 32 (1971).
[16] J. N. Baptist, R. K. Gholson and M. J. Coon, Biochim. Biophys. Acta 69, 40 (1963).
[17] see Chapter 10 in ref. [4].
[18] J. E. Potts, R. A. Clendinning, W. B. Ackart and W. D. Niegisch, "The Biodegradability of Synthetic Polymers" in J. Guillet (ed.), "Polymers and Ecological Problems", Plenum Press, New York (1973).
[19] J. E. Potts, R. A. Clendinning and W. B. Ackart, U.S.E.P.A.Contract CPE-70-124 (1972); see also ref. [2].

[20] *A. Schwartz*, "Mikrobielle Korrosion von Kunststoffen und ihren Bestandteilen", Abhandl. deutsch. Akad. Wiss. Berlin, Kl. Chem. Geol. u. Biol. 5, 1 (1963).

[21] *N. B. Nykvist*, Plastics Polymers 42, 195 (1974).

[22] *A.-C. Albertson* and *B. Rånby*, ref. [3], p. 743.

[23] *A.-C. Albertson, Z. G. Banhidi* and *L. L. Beyer-Erikson*, J. Appl. Polym. Sci. 22, 3419 and 3435 (1978).

[24] *L. Jen-Hao* and *A. Schwartz*, Kunststoffe 51, 317 (1961).

[25] *J. Trauter*, Chemiefasern/Textilindustrie 26, 349 (1976).

[26] *J. P. Casey* and *D. G. Manly*, ref. [3], p. 819 and references cited therein.

[27] *R. Zahn* and *H. Wellens*, Chemikerzeitung, 98, 228 (1974).

[28] *H. Jalke*, "Entschlichtungsmittel" in *A. Chwala* and *V. Anger* (eds.), "Handbuch der Textil-hilfsmittel", Verlag Chemie, Weinheim (1977).

[29] *W. J. Bailey, Y. Okamato, W.-C. Kuo* and *T. Narita*, ref. [3], p. 765.

[30] *R. D. Fields, F. Rodriguez* and *R. K. Finn*, J. Appl. Polym. Sci. 18, 3571 (1974).

[31] *E. J. Frazza* and *E. E. Schmitt*, J. Biomed. Mater. Res. Symp. 1, 43 (1970).

[32] *S. J. Huang, M. Bitritto, G. Brenkle, K. W. Leong, J. A. Pavlisko, M. S. Roby, J. P. Bell, J. A. Cameron* and *J. R. Knox*, Am. Chem. Soc. Div. Polym. Chem., Polym. Preprints 18, 438 (1977).

[33] *R. D. Fields* and *F. Rodriguez*, ref. [3], p. 775.

[34] *S. J. Huang, J. P. Bell, J. R. Knox, H. Atwood, D. Bansleben, M. Bitritto, W. Borghard, T. Chapin, K. W. Leong, K. Natarjan, J. Nepumuceno, M. Roby, J. Soboslai* and *N. Shoemaker*, ref. [3], p. 731.

[35] *H. W. Steinmann*, Polym. Preprints 11, 285 (1970); U.S. Patent 3,386,932.

[36] *S. Kim, V. T. Stannett* and *R. D. Gilbert*, J. Macromol. Sci. Chem. A 10, 671 (1976); J. Polym. Sci. Polym. Lett. Ed. 11, 731 (1973).

[37] *R. J. Dennenberg, R. J. Bothast* and *T. P. Abbott*, J. Appl. Polym. Sci. 22, 459 (1978).

[38] *K. H. Wallhäusser*,
(a) Conference on Degradability of Polymers and Plastics, London (1973), paper 17;
(b) Müll und Abfall 4, 10 (1972);
(c) Kunststoffe 63, 54 (1973); Verpack.-Rundsch. 23, 266 (1972).

[39] *E. Spohn*, "Neuere Verfahren der Kompostierung. Prospekt des Instituts für Bodenhygiene, Blaubeuren (1970).

[40] *A. Tsugita*, "Phage Lysozyme and Other Lytic Enzymes", in Vol. 5 of ref. 9 (a).

7   Chemical Degradation

7.1   Introduction

The last chapter of this book is devoted to chemical reactions which start spontaneously when certain low molecular weight compounds are brought in contact with polymers. Commonly, the rate of chemical reactions is strongly dependent on temperature, which implies that thermal and chemical processes overlap. Thermal degradation can be distinguished from chemical degradation, however, by restricting the former term to those processes which are initiated solely by heating, i.e. without the addition of another compound. Admittedly, this definition is somewhat arbitrary insofar as really pure polymers do not exist. In practice, all polymers contain small amounts of low molecular weight compounds (impurities, plasticizers etc.) which might undergo chemical reactions with the macromolecules as the temperature is increased. However, in this chapter only processes are considered which occur upon addition of low molecular weight compounds to macromolecules. Chemical reactions which are initiated by heating "pure" polymers have been dealt with in Chapter 2 (Thermal Degradation). Certain limitations are necessary because of the vast number of possible reactions and, therefore in accordance with the scope of this book, only those chemical reactions leading to main-chain scission and/or deterioration of physical properties will be treated. Consequently, emphasis will be given to aspects concerning the stability of polymers against chemicals. In this respect synthetic and natural polymers containing hydrolyzable linkages will receive major attention. Polymers containing unsaturated carbon-carbon bonds are also of interest, because they are susceptible to metathesis reactions or reactions with ozone.

Reactions of polymers with molecular oxygen are of general importance because oxygen is ubiquitous. Therefore, oxidative processes cause problems not only in outdoor exposures of plastics (weathering) but also in processing. The latter aspect has been referred to in Chapter 2. Frequently, oxidative degradation proceeds according to free radical chain reactions (autoxidation, see Chapter 1) initiated by UV-light, γ-radiation, mechanical stress etc. Chemical modes of free radical initiation exist also and will be dealt with below.

Recently, because air pollution has become a worldwide problem, the behavior of plastics towards pollutant gases (NO_2 and SO_2) has gained importance and has become the subject of investigations.

Last but not least, the importance of the stability of polymers against solvents has to be considered. Frequently, solid polymeric articles change or lose shape or deteriorate on immersion into liquid chemicals. This phenomenon might be caused by chemical reactions or, on the other hand, be the result of a physical process consisting of a strong interaction between the polymer and the liquid. Under the influence of this interaction, the polymer dissolves eventually or swells to a limited degree but remains chemically intact. From the practical point of view, the effect of chemical or physical interaction on the properties of the polymer is equivalent. The term "chemical stability" is, therefore, commonly used in a broad sense covering both chemical and physical interaction of low molecular weight compounds with polymers.

Another interesting phenomenon concerns the fact that many polymers consist of amorphous and crystalline fractions. For certain chemical reactions, e.g. hydrolysis, the crystalline regions have been found impervious to chemical agents. Thus, the amorphous fraction of the polymer is attacked selectively, a fact capable of being utilized, for instance, for analytical purposes (determination of the degree of crystallinity).

In the following sections of this chapter, selected topics of chemical degradation will be treated in some detail. The selection has to be arbitrary, of course. It might, however, lead to a comprehension of the problems arising when polymers come in contact (whether on purpose or not) with aggressive chemicals.

Readers seeking additional information on chemical degradation of polymers are referred to various books and articles [1—15].

7.2 Solvolysis

Generally, solvolysis reactions concern the breaking of C—X bonds, X designating hetero (non-carbon) atoms, i.e., in the context of this book: O, N, P, S, Si or halogen. Of primary interest are solvolysis reactions of polymers containing hetero atoms in the main chain, because in these cases, solvolysis implies main-chain rupture as indicated by reaction (7.1):

$$
\begin{array}{c}
-\overset{\displaystyle\rangle}{\underset{\displaystyle X}{C}}- \\
\underset{\displaystyle\langle}{\overset{\displaystyle|}{X}} \\
-\overset{\displaystyle|}{\underset{\displaystyle\rangle}{C}}-
\end{array}
\;+\; \text{YZ} \;\longrightarrow\;
\begin{array}{c}
-\overset{\displaystyle\rangle}{\underset{\displaystyle X}{C}}- \\
\underset{\displaystyle Z}{\overset{\displaystyle|}{X}}
\end{array}
\;+\;
-\overset{\displaystyle Y}{\underset{\displaystyle\langle}{C}}-
\qquad (7.1)
$$

Common solvolysis agents (YZ) are water, alcohols, ammonia, hydrazine etc.

Regarding polymer degradation hydrolysis (YZ = HO—H) has received prominence. Table 7.1 lists typical examples. Water soluble polymers (e.g. most polysaccharides) are rather readily hydrolyzed. On the other hand, water insoluble polymers are attacked only very slowly. In these cases the occurrence of the reaction is restricted to the surface of the specimen and the ability of the polymer to adsorb water plays an important role. Crystallinity and chain conformation exert a strong influence, as will be outlined below.

Generally, the mechanism prevailing in neutral or acidic media differs from that in alkaline media, as will be demonstrated here for ester linkages. At pH $\leq$ 7, hydrolysis is initiated by a protonation process which is followed by the addition of H_2O and the cleavage of the ester linkage. This is illustrated in a simplified manner by reactions (7.2) and (7.3):

$$
\sim\!\!\overset{\displaystyle O}{\overset{\displaystyle\|}{C}}\!\!-\!\!O\!\!\sim \;+\; H^{\oplus} \;\longrightarrow\;
\left[\;
\sim\!\!\overset{\displaystyle\overset{\oplus}{O}H}{\overset{\displaystyle\|}{C}}\!\!-\!\!O\!\!\sim \;\leftrightarrow\;
\sim\!\!\overset{\displaystyle OH}{\underset{\displaystyle\oplus}{C}}\!\!-\!\!O\!\!\sim
\;\right]
\qquad (7.2)
$$

$$
\sim\!\!\overset{\displaystyle OH}{\underset{\displaystyle\oplus}{C}}\!\!-\!\!O\!\!\sim \;+\; H_2O \;\rightleftharpoons\;
\sim\!\!\overset{\displaystyle OH}{C}\!\!=\!\!O \;+\; H^{\oplus} \;+\; HO\!\!\sim
\qquad (7.3)
$$

Table 7.1   Hydrolysis of linear polymers (typical examples)

main-chain linkage under attack	products of hydrolysis	examples
$-\overset{\mid}{\underset{\mid}{C}}-\overset{\mid}{\underset{\parallel}{C}}-O-\overset{\mid}{\underset{\mid}{C}}-$ O carboxylic acid ester	$-\overset{\mid}{\underset{\parallel}{C}}-OH + HO-\overset{\mid}{\underset{\mid}{C}}-$ O	polyester
$-O-\overset{O^{(-)}}{\underset{\parallel}{\overset{\mid}{P}}}-O-\overset{\mid}{\underset{\mid}{C}}-$ O phosphoric acid ester	$-O-\overset{O^{(-)}}{\underset{\parallel}{\overset{\mid}{P}}}-OH + HO-\overset{\mid}{\underset{\mid}{C}}-$ O	nucleic acids (DNA etc.)
$-\overset{\mid}{\underset{\mid}{C}}-O-\overset{\mid}{\underset{\mid}{C}}-$ ether, glycoside	$-\overset{\mid}{\underset{\mid}{C}}-OH + HO-\overset{\mid}{\underset{\mid}{C}}-$	polyether, polysaccharides (cellulose, amylose etc.)
$-\overset{\mid}{\underset{\mid}{C}}-\overset{\mid}{\underset{O}{\overset{\parallel}{C}}}-\overset{H}{\underset{\mid}{N}}-\overset{\mid}{\underset{\mid}{C}}-$ amide (peptide)	$-\overset{\mid}{\underset{\mid}{C}}-\overset{\mid}{\underset{O}{\overset{\parallel}{C}}}-OH + H_2N-\overset{\mid}{\underset{\mid}{C}}-$	polyamides, proteins, polypeptides
$-\overset{\mid}{\underset{\mid}{C}}-O-\overset{\mid}{\underset{O}{\overset{\parallel}{C}}}-\overset{H}{\underset{\mid}{N}}-\overset{\mid}{\underset{\mid}{C}}-$ urethane	$-\overset{\mid}{\underset{\mid}{C}}-OH + CO_2 + H_2N-\overset{\mid}{\underset{\mid}{C}}-$	polyurethanes
$-\overset{\mid}{\underset{\mid}{Si}}-O-\overset{\mid}{\underset{\mid}{Si}}-$ siloxane	$-\overset{\mid}{\underset{\mid}{Si}}-OH + HO-\overset{\mid}{\underset{\mid}{Si}}-$	polydialkylsiloxanes

In alkaline media hydroxyl ions are attached to the carbonyl carbons. Subsequently ester linkages are ruptured, as illustrated for the case of a polyester by the following reactions:

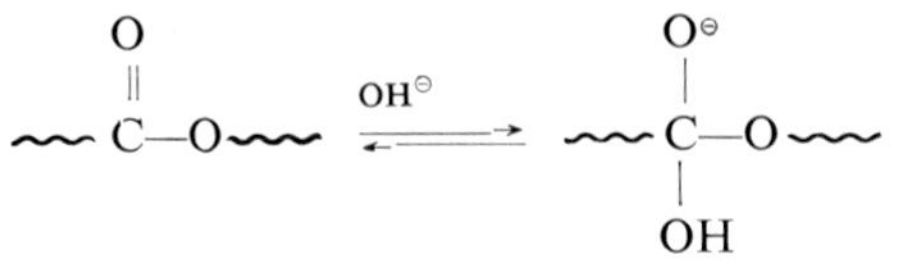

$$\sim\!\!\sim\!\! \overset{O}{\overset{\parallel}{C}}\!-\!O\!\sim\!\!\sim \underset{\longleftarrow}{\overset{OH^{\ominus}}{\longrightarrow}} \sim\!\!\sim\!\! \underset{\underset{OH}{\mid}}{\overset{\overset{O^{\ominus}}{\mid}}{C}}\!-\!O\!\sim\!\!\sim \tag{7.4}$$

$$\begin{array}{c}\sim\!\!\overset{O^{\ominus}}{\underset{\underset{OH}{|}}{\overset{|}{C}}}\!\!-\!O\!\sim\end{array}\ \rightleftharpoons\ \begin{array}{c}\sim\!\!C\!\overset{O}{\overset{\|}{}}\!\!-\!OH\ +\ {}^{\ominus}O\!\sim\\[4pt]\downarrow\\[4pt]\sim\!\!C\!\overset{O}{\overset{\|}{}}\!\!-\!O^{\ominus}\ +\ HO\!\sim\end{array}\qquad(7.5)$$

More complicated mechanisms are expected for the alkaline degradation of other polymers. It has been pointed out by *Vollmert* [89], for instance, that in the case of polysaccharides (cellulose etc.) alkaline hydrolysis is preceded by an oxidative process generating carbonyl groups on the glucose units.

In connection with solvolysis, the reader's attention is drawn to hydrolytic degradation processes induced by microbial attack of polmers (see Section 6.3). In this case, hydrolytic cleavage is accomplished commonly by enzymatic action. There are, however, cases where microbial metabolism causes a pH change in the environment of polymers with the consequence of hydrolytic cleavage of sensitive linkages in the main chains or in side groups.

There are three interesting aspects concerning solvolysis reactions which will be discussed below in some detail: characterization of polymers, stability of plastic articles and commercial applications (including recycling).

Additional information concerning solvolytic degradation will be presented in Section 7.6 in connection with the discussion of selective degradation processes.

7.2.1 Polymer Characterization

A classical method for the characterization of organic compounds consists in specifically converting or splitting off characteristic groups with the aim of obtaining simple compounds which can be readily identified and which are amenable to quantitative determination. Solvolysis is an important standard method which has been applied successfully to polymer characterization since the very beginning of systematic studies in the field of macromolecules. The development of the method received impetus by the advent of new elegant analytical separation methods, mainly the various modes of chromatography, in combination with modern detection techniques, such as mass spectrometry.

The aim of solvolytic techniques is to cleave all susceptible linkages thus fragmenting the macromolecules into their components and so allowing the quantitative determination of the composition of the polymer. Relevant applications include the determination of the composition of polyesters, alkyd resins or polyurethanes. A polyurethane, for example, might be composed of 4,4'-diphenylmethane diisocyanate, poly(ethylene glycol adipate) and 1,4-butanediol. The content of adipic acid, ethylene glycol and 1,4-butanediol can be determined after solvolysis. Another example pertains to alkyd resins being composed of glycerine and various dicarboxylic acids. Solvolysis methods frequently used are listed in Table 7.2. Typical gas chromatograms identifying the components of an alkyd resin and a polyester are shown in Fig. 7.1.

In most cases drastic conditions are required (elevated temperatures and/or long reaction times). The search for mild reaction conditions revealed that tetramethylammonium hydroxide (TMAH, $[(CH_3)_4N]OH$) reacts very fast at relatively low temperatures with

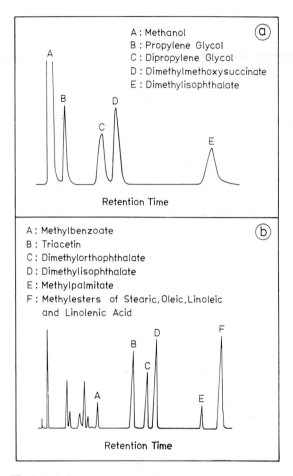

Fig. 7.1   Polymer characterization by solvolysis.
Gas chromatograms representing components of a polyester (a) [21] and an alkyd resin (b) [20]. Prior to analysis the polymers were decomposed by methanolysis with potassium methoxide (a) and with lithium methoxide (b)

polyesters, polyurethanes, alkyd resins, and certain polyacrylates (e.g. butyl and ethyl esters) [16–18]. According to *West* [16], the substrate is initially saponified by refluxing with TMAH in methanol for periods up to 15 min and subsequently esterified with methyl iodide in the presence of N,N-dimethylformamide/methanol solvent. Recently, it was reported that in the case of polyesters, the methanolysis is already accomplished in the first stage [18]:

$$\text{R—C—OR}' + \text{CH}_3\text{OH} \xrightarrow{\text{TMAH}} \text{R—C—OCH}_3 + \text{R}'\text{OH} \tag{7.6}$$
$$\overset{\parallel}{\text{O}} \qquad\qquad\qquad\qquad \overset{\parallel}{\text{O}}$$

Table 7.2 Polymer characterization via solvolysis
(adopted from *Mosselman* and *de Wit* [18])

Polymer	Method	Ref.
polyesters	methanolysis with TMAH *)	16, 17, 18
	methanolysis with alkalimethoxides	20, 21
	methanolysis with BF$_3$	20
	aminolysis	19
	hydrazinolysis	19
polyamides	acid hydrolysis	22
polyurethanes	methanolysis with TMAH *)	18
	alkaline hydrolysis	23
alkyd resins	methanolysis with TMAH *)	16
	methanolysis with LiOCH$_3$	8
polyacrylates	methanolysis with TMAH *)	18
	hydrolysis with KOCH$_3$ in methanol	25
polymethacrylates	methanolysis with KOCH$_3$	18
	(alkaline fusion)	26

*) TMAH: tetramethylammonium hydroxide [(CH$_3$)$_4$N]OH

7.2.2 Stability of Polymeric Materials

The stability of polymeric materials against solvolytic agents is of great importance for a variety of applications. Frequently, polymeric materials, e.g. plastic containers or tubings, are exposed over extended periods to the influence of acidic or alkaline reagents. As has been mentioned before, polymers containing hydrolyzable linkages, such as polyesters, polyurethanes and polyamides are prone to the attack by acids and bases. On the other hand, polyhydrocarbons, e.g. polyethylene and polypropylene, are quite stable. Moreover, crystallinity improves the stability, because it makes the material impervious to aggressive agents. Commonly, solvolysis reactions require a significant activation energy implying an increase of the solvolysis rate with increasing temperature, i.e. a diminution of the stability with increasing temperature.

In Table 7.3, typical polymers are listed together with the stability characterizations "satisfactory" ($+$) or "unsatisfactory" ($-$), indicating whether a polymer can be utilized in an acidic or basic medium or not. Problems concerning the classification of polymers according to their stability will be discussed again in Section 7.7. It appears that available standard tests [27] do not take into account modern technological demands. For example, these tests do not measure the long term stability of polymers against attack by chemical reagents or the dependence of long term stability on mechanical stress and temperature. Standard testing procedures to measure these parameters have been proposed [28], however. The need for such tests is based on the fact that the resistance of a polymer against solvolytic attack is dependent on variations in temperature and mechanical stress. An increase of temperature can cause melting of crystallites, thus making larger portions of the polymer amenable to the attack of solvolytic agents. Moreover, if the polymer comes in contact with oxidizing agents (HNO_3, H_2CrO_4 etc.), oxidative reactions might cause additional deterioration at elevated temperatures.

Table 7.3   Stability of selected polymers against solvolytic agents at ambient temperature

polymer	acidic media	alkaline media*)
polyethylene	+	+
polypropylene	+	+
poly-1-butene	+	+
polyisobutene	+	+
polystyrene	+	+
polytetrafluoroethylene	+	+
polytrifluorochloroethylene	+	+
polyvinylfluoride	+	+
polyvinylchloride (unplasticized)	+	+
polyvinylchloride (plasticized)	−	−
polymethylmethacrylate	−	−
polyacrylonitrile	−	−
polyoxymethylene	−	−
polysulfones	−	−
polyamides	−	−
polycarbonates	−	−
polyethyleneterephthalate	−	−
polyurethanes	−	−
phenolformaldehyde resins	+	−
polydimethylsiloxanes	−	−
natural rubber	+	+
butyl rubber	+	+
unsaturated polyesters (UPE)	+	−

*) +: satisfactory; −: unsatisfactory

7.2.3   Commercial Applications Including Recycling

On a commercial scale, hydrolytic processes serve to produce low molecular weight sugars from polysaccharides (starch, cellulose) [29]. Furthermore, hydrolysis is of technical importance for the decomposition of starch used as sizing agent in the textile industry [30].

Because of shortages in raw materials (crude oil) for the synthesis of plastics, recycling of waste plastics has become very important recently. Due to the incompatibility of most polymers and the enormous difficulties encountered in segregating polymeric

articles of different chemical composition found in plastics waste, complete recycling is difficult and limited to special cases. Therefore, hydrolytic processes are being developed for the utilization of plastics scrap and waste as either an energy source or as raw material for the production of valuable chemicals. (Corresponding industrial processes based on thermolysis and pyrolysis were discussed in Section 2.8).
Some of the various chemical processes developed so far are compiled in Table 7.4 [31]. It appears that most processes are rather complicated, apart from the methanolysis of polyethyleneterephthalate which yields the starting products for the synthesis of the

Table 7.4   Utilization of scrap and waste plastics as raw materials for the production of chemicals (according to *J. Brandrup* [31])

polymer	procedure	products	company
polyolefins	$120-140\,°C$, O_2	oxidized waxes	Hoechst AG[a]
	$350-500\,°C$ H_2, $ZnCl_2$ (catalyst)	low molecular weight hydrocarbons (gasoline)	AG. Ind. Sci. Techn./Japan
polystyrene	$250-500\,°C$ $H_2/ZnCl_2/$ $Al_2O_3-SiO_2$	ethylbenzene	AG. Ind. Sci.[b] Techn./Japan
	$300\,°C$, H_2O, CuO	styrene	Shimada/Japan Yamaguchi/Japan
polyvinyl-chloride	heat, Cl_2	carbontetrachloride	Hoechst AG
	1st stage $500\,°C$, SO_2 2nd stage $200\,°C$, Cu(I)	dichloroethane	Asahi/Japan[c]
polyethylene-terephthalate	heat, methanol	dimethylterephthalate, ethyleneglycol	Hoechst AG[d]
poly-urethanes	heat, H_2O	polyetherglycols, tolylenediamine	Bayer AG[e] Ford General Motors Upjohn
polyamides	$250-300\,°C$ NH_3, H_2	diamines	ICI[f]

(a) Hoechst, DOS 2035706 (1970)
(b) Agency Ind. Sci. Technol., JA 4843482 (1971)
(c) Asahi Chemical Ind. Co. JA J 7 4029-163 (1970)
(d) Hoechst, DOS 1495251 (1962)
(e) Bayer AG, in connection with "Programm des Verbandes der kunststofferzeugenden Industrie". See also: Chem. Ind., Düsseldorf 26, 24 (1974),
    E. Grigat, Kunststoffe 68, 281 (1978)
(f) ICI, NL 6709663 (1967)

polymer (in analogy to the thermolysis of polymethylmethacrylate). Other interesting industrial applications are: the generation of oxidized waxes or of gasoline, by reacting polyolefins with O_2 or H_2, respectively, the production of ethylbenzene from polystyrene, and the decomposition of polyvinylchloride into CCl_4 in the presence of Cl_2.

To the best knowledge of the author, however, only a few processes have been developed to the stage of commercial application so far, such as the methanolysis of polyethylene-terephthalate (Hoechst) and the hydrolysis of polyurethanes (Bayer) [134].

7.3   Reactions of Olefinic Double Bonds

There exist a limited number of important natural and synthetic homopolymers containing C—C double bonds in the main chain. Typical examples are natural rubber (poly-cis-1,4-isoprene) and poly-1,4-butadiene, i.e. polymers possessing one double bond per repeating unit. Moreover, there exist a number of technically interesting copolymers containing olefinic comonomers. Also certain commercial polymers may contain carbon double bonds as a "chemical impurity". As well known, C—C double bonds are reactive towards various chemicals. Metathesis and ozonization are two chemical reactions of olefinic double bonds leading to the scission of the C—C double bonds. They have received prominence mainly for their utilization in the analysis of the chemical structure of relevant homo- and copolymers.

7.3.1   Metathesis

Metathesis is a transition metal catalyzed reaction involving olefinic compounds [32—34] and proceeding as depicted generally in Scheme 7.1.

Scheme 7.1   General scheme of metathesis. "Me" denotes a catalyst system consisting of a catalyst/co-catalyst pair (e.g. $WCl_6/(CH_3)_4Sn$)

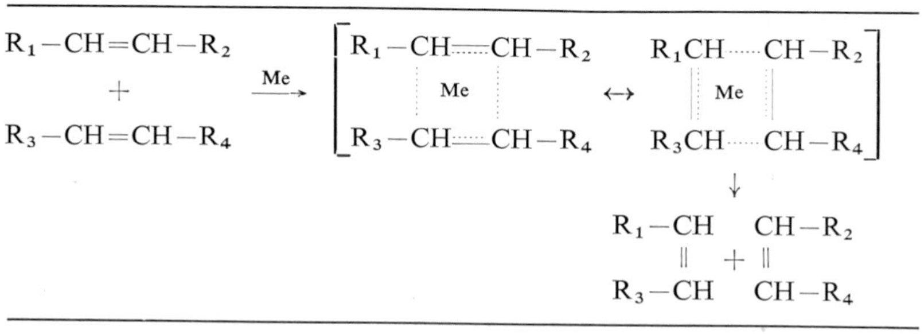

$$R_1-CH=CH-R_2$$
$$+ \xrightarrow{Me} \left[\begin{array}{ccc} R_1-CH\cdots\cdots CH-R_2 & & R_1CH\cdots\cdots CH-R_2 \\ | \quad Me \quad | & \leftrightarrow & | \quad Me \quad | \\ R_3-CH\cdots\cdots CH-R_4 & & R_3CH\cdots\cdots CH-R_4 \end{array} \right]$$
$$R_3-CH=CH-R_4$$

$$\downarrow$$

$$R_1-CH \quad CH-R_2$$
$$\| \; + \; \|$$
$$R_3-CH \quad CH-R_4$$

Metathesis requires the presence of appropriate catalyst/co-catalyst pairs (catalysts: WCl_6, $MoCl_5$, $ReCl_5$ etc., co-catalysts: $C_2H_5AlCl_2$, $Sn(CH_3)_4$, $Sn(C_6H_5)_4$, 1,2-divinyl-cyclohexane etc.). In polymer chemistry, metathesis is of great importance as, apart from other interesting applications, especially in the synthesis of polymers (ring opening polymerization) [35—36], it can be used to break down homo- and copolymers containing olefinic unsaturations in the main chains. In this latter case, the polymer is reacted with a low molecular weight olefinic compound as indicated by reaction (7.7):

$$
\begin{array}{ccccc}
\sim\!\!\sim \text{CH}=\text{CH} \sim\!\!\sim & & \sim\!\!\sim\text{CH} & \text{CH} \sim\!\!\sim & \\
+ & \rightarrow & \| & + & \| & \qquad\qquad (7.7)\\
\text{R}_1-\text{CH}=\text{CH}-\text{R}_2 & & \text{R}_1-\text{CH} & \text{CH}-\text{R}_2 &
\end{array}
$$

Intense work has been devoted to the application of metathesis to various problems in polymer chemistry [36—42]. Of special interest is work concerning the determination of length distributions of sequences in copolymers [37—40] e.g. in partially hydrogenated poly-1,4-butadiene, or work concerning the determination of the content of polyethylene in crosslinked mixtures of poly-1,4-butadiene and polyethylene [42]. Commonly, polymer fragments generated by reaction (7.7) are separated for analysis by appropriate techniques such as gel permeation chromatography or gas chromatography. Metathesis reactions are possible also in homopolymers possessing a sufficiently high degree of unsaturation. With polypentenamer, polybutadiene and polyisoprene, for instance, cyclic structures (e.g. cyclopentene) are formed upon intramolecular metathesis degradation. The analogous intermolecular reaction causes a reorganization of the molecular weight distribution [36, 41].

7.3.2   Ozonization

Ozone is known to react with most organic compounds. The rate of reaction with saturated materials is rather low and, therefore, of minor importance for straightforward degradation processes. However, reactions of ozone with saturated compounds might serve to initiate autoxidation processes (see below).

Contrary to the reaction with saturated compounds, O_3 reacts readily with olefinic double bonds causing the scissioning of these bonds, a process denoted as "ozonolysis". Generally, the reaction of O_3 with olefinic double bonds in polymers proceeds quite readily as long as the double bonds are accessible. Since ozone is electrophilic, its rate of reaction with a double bond increases if that bond is substituted with an electron donating group and, vice versa, the rate decreases if the substituent is an electron acceptor. Thus, polychloroprene is more resistant against ozone than polyisoprene.

For practical applications of rubber and similar materials, the ozone-induced formation of cracks causes problems. The generation of ozonization cracks requires, however, a minimum mechanical stress, otherwise the reaction is restricted to the specimen's surface [43]. After ozone treatment of compact samples of rubber, polybutadiene etc., carbonyl and carboxyl groups as well as crosslinked regions are detectable at the surface. In solution main-chain degradation predominates. If molecular oxygen is present, O_3 treatment initiates autoxidation.

Frequently discussed mechanisms [44—46] concerning the reaction of O_3 with olefinic double bonds are based on the idea that five-membered cyclic intermediates (I and II) are formed:

According to *Benson* [44], I decomposes into a carbonyl and a biradical:

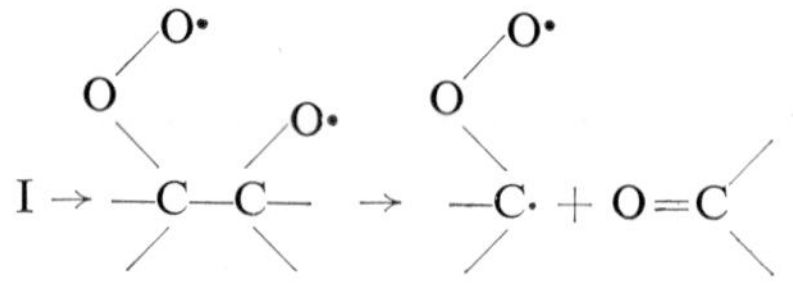

$$I \rightarrow -\underset{|}{C}-\underset{|}{C}- \rightarrow -\overset{|}{C}\cdot + O=C \tag{7.8}$$

Criegee [45] assumed that zwitterion/ketone pairs are formed intermittently. In many cases these species recombine to II:

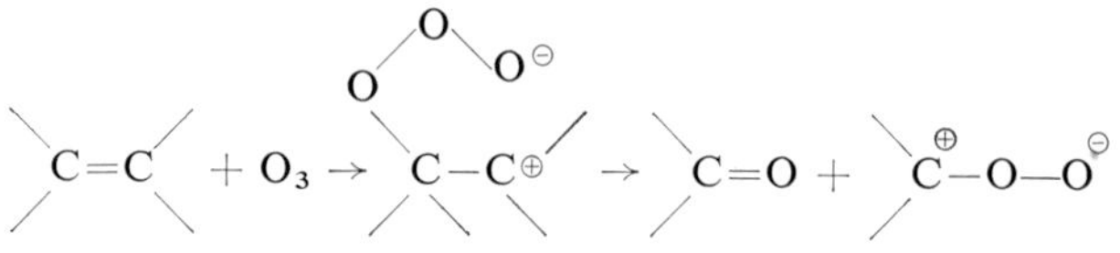

$$C=C + O_3 \rightarrow C-C^\oplus \rightarrow C=O + \overset{\oplus}{C}-O-O^\ominus \tag{7.9}$$

$$C=O + \overset{\oplus}{C}-O-O^\ominus \rightarrow II \tag{7.10}$$

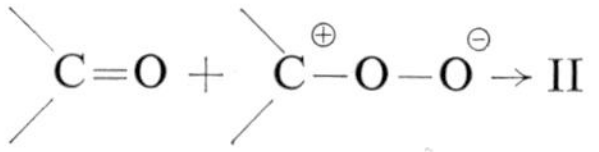

II decomposes at a later stage into free radicals. Therefore, no matter which mechanism prevails, free radicals are presumed to be formed eventually. They are capable of abstracting hydrogen atoms from neighboring base units (either intra- or intermolecularly). Thus, in the presence of O_2, the initiation of autoxidation processes (see Chapter 1) becomes feasible.

The possibility of O_3 directly attacking saturated hydrocarbons at ambient temperature, has been occasionally investigated [47—51]. Substances used for these studies comprise cis- and trans-decalin, adamantane, cyclohexane etc. A proposed mechanism is illustrated in Scheme 7.2.

Scheme 7.2   Mechanism of the direct attack of saturated hydrocarbons by ozone

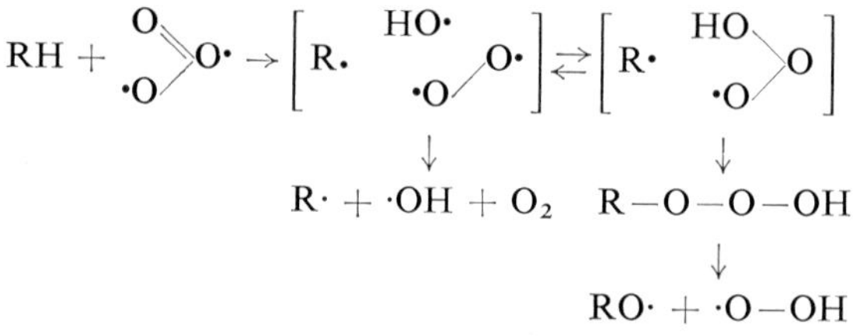

$$RH + \overset{O}{\underset{\cdot O}{\diagup}}O\cdot \rightarrow \left[R\cdot \overset{HO\cdot}{\underset{\cdot O}{\diagup}}O\cdot \right] \rightleftharpoons \left[R\cdot \overset{HO}{\underset{\cdot O}{\diagup}}O \right]$$

$$\downarrow \qquad\qquad \downarrow$$

$$R\cdot + \cdot OH + O_2 \qquad R-O-O-OH$$

$$\downarrow$$

$$RO\cdot + \cdot O-OH$$

It should be pointed out that reactive free radicals (·OH, RO·, ·OOH) are produced according to this mechanism. These radicals will abstract hydrogen atoms from surrounding hydrocarbon molecules:

$$\cdot OH + RH \rightarrow H_2O + R\cdot \tag{7.11}$$

$$\cdot OOH + RH \rightarrow H_2O_2 + R\cdot \tag{7.12}$$

$$RO\cdot + RH \rightarrow ROH + R\cdot \tag{7.13}$$

In the presence of molecular oxygen, radicals R· will readily undergo the reaction

$$R\cdot + O_2 \rightarrow R-O-O\cdot \tag{7.14}$$

and, therefore, the direct attack of saturated hydrocarbons by O_3 implies the initiation of autoxidation.

Many polymers, especially polyolefins, contain a small amount of unsaturation. Therefore, it is difficult to assess the role of the reaction of O_3 with saturated structures in those polymers. In a recent study concerning the effect of ozone on the degradation of polyvinylchloride (PVC), the influence of ozone on the formation of HCl and of peroxidic groups was investigated [52]. As can be seen from Fig. 7.2 (a) ozone markedly

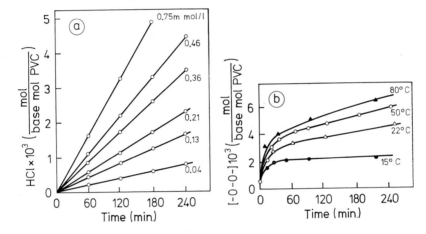

Fig. 7.2   The effect of ozone on the degradation of powdered polyvinylchloride (PVC). (a) Elimination of HCl from PVC at 80 °C in the presence of O_2 containing O_3 at concentrations given in the graph.
(b) Formation of peroxide groups in the presence of O_2 containing 4.6×10^{-4} mol/l O_3 at various temperatures, as indicated in the graph (according to *Abdullin* et al. [52])

accelerated the elimination of HCl during the treatment of powdered PVC with O_3-containing oxygen (at 80 °C). Fig. 7.2 (b) shows the increase of the content of peroxide groups in PVC as a function of time at various temperatures. The two branches of the curves in Fig. 7.2 (b) suggest the occurrence of two processes: a rapid process due to the reaction of O_3 with C=C bonds and a slow process due to the reaction of ozone with C—H bonds. The rate constants of the two reactions differ by a factor of 10^6 between 0 and 20 °C and depend on temperature according to Eq. (7.15) and (7.16) [52]:

$$k_{O_3+C=C} = (10 \pm 2) \, 10^6 \exp\left(-\frac{3500 \pm 500}{RT}\right) \qquad (7.15)$$

$$k_{O_3+CH} = (9 \pm 2) \, 10^8 \exp\left(-\frac{13500 \pm 1500}{RT}\right) \qquad (7.16)$$

(rate constants in l/mol s and activation energies in cal/mol degree)

Table 7.5   Ozonization of water-soluble polymers
(adopted from *Suzuki* et al. [53])

polymer	ozonization time (h)	ozone consumed (mg/g polymer)	MW
poly(ethylene oxide)	0	0	8 000
	2	836	250
polyvinylalcohol	0	0	28 000
	4	368	460
polyvinylpyrrolidone	0	0	27 000
	4	1 273	560
sodium polyacrylate	0	0	410 000
	4	860	250
polyacrylamide	0	0	280 000
	4	910	340

Practical interest in ozone treatment of various water-soluble polymers developed recently with the aim of improving their biodegradability [53]. The polymers compiled in Table 7.5 were reportedly released into waste water canals leading to rivers and lakes after having been used in industrial processes. As discussed in Section 6.4 only polyvinyl alcohol is relatively readily biodegradable. From Table 7.5 it can be inferred that O_3 treatment reduces the average MW in all cases, leading to a significant improvement in biodegradability except in the case of polyacrylamide.*) For poly(ethylene oxide) the mechanisms shown in Scheme 7.3 were proposed [54]. It was taken into account that the polymer is randomly fragmented by ozone, that the fragments contain terminal OH-groups and that, moreover, formic esters are formed as a major product. It is assumed that ether bonds undergo an electrophilic attack by ozone as proposed for the ozonization of simple ethers [55].

For further reading, concerning the reaction of ozone with various polymers, articles by *Jellinek* [56] and *Murray* [46] (and literature references therein) may be referred to.

Since ozone severely affects certain polymers, especially elastomers containing unsaturations, testing procedures for the ozone resistance of polymers have been developed and adopted as standard testing methods [57, 58].

---

*) With polyacrylamide the rate of biodegradation was found to be independent of the average MW.

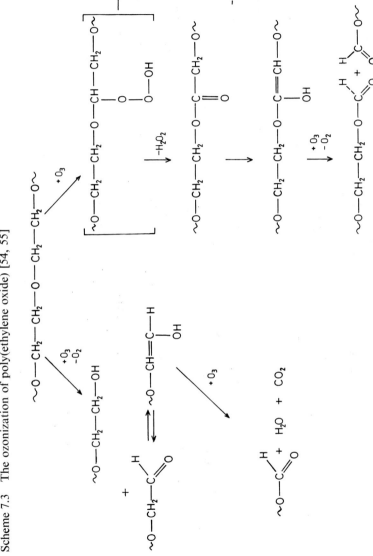

Scheme 7.3   The ozonization of poly(ethylene oxide) [54, 55]

7.4   Oxidative Degradation

7.4.1   General Aspects

From a practical point of view, it seems to be appropriate to distinguish between "direct" oxidation of polymers with certain compounds and autoxidation of polymers. The term direct oxidation refers here to reactions proceeding under mild conditions, i.e. to reactions occurring spontaneously at ambient or slightly increased temperatures and proceeding kinetically as one-step-reactions. Typical examples are the oxidation of hydrocarbon polymers with $KMnO_4/H_2SO_4$ or nitric acid (HNO_3), the oxidation of starch with sodium bromite ($NaBrO_2$) and sodium persulfate ($Na_2S_2O_8$), and the oxidation of various polymers possessing functional groups with metal ions (e.g. of Cu, Co, Ni, Mn). Oxidations play an important role in the characterization of partially crystalline polymers if the amorphous material is preferentially oxidized, thus allowing the separation of crystallites (see Section 7.6). $NaBrO_2$ and $Na_2S_2O_8$ serve as oxidizing agents for the removal of starch from cellulosic fibers, when starch is used as sizing agent [30, 59]. Oxidation with metal ions might be hazardous for the stability of polymers, if the mechanical properties of the polymers are deteriorated by this process. Deterioration is commonly observed if autoxidation occurs. (As elucidated in Chapter 1, autoxidation refers to reactions of materials with molecular oxygen which proceed as chain reactions.)

Recent articles [60, 61] reviewing the field of oxidative degradation stressed autoxidation and neglected direct oxidation processes. This appears to be justified, since direct oxidation processes have been extensively studied with low molecular weight compounds and have been dealt with in organic chemistry textbooks and in books devoted in particular to this subject [62].

Owing to the economical importance of polymer stability, autoxidation has been emphasized by all those concerned with the application of plastics and detailed treatments of the subject are available [13, 60, 61, 63]. Therefore, it seems appropriate to concentrate here only on a few aspects of autoxidation which are of interest in the context of this book.

7.4.2   Modes of Initiation of Autoxidative Processes

At ambient temperatures the chemical structures of most polymers are quite stable against the attack of molecular oxygen (O_2). That oxidative degradation is a quite common phenomenon notwithstanding, is due to the ease with which various initiating reactions occur. According to the present state of knowledge, the most prominent modes of initiation refer to the generation of free radicals capable of reacting rapidly with O_2. As has been outlined in the preceding chapters of this book, these modes comprise thermolysis (pyrolysis), photolysis, radiolysis and mechanical stress-induced reactions. In addition, free radical formation is feasible in processes pertaining to electrical phenomena such as corona discharge, electrical breakdown and electrolysis. Since electrical processes are only rarely becoming important for oxidative processes in polymers, they are not discussed in detail in this book.

Chemical modes of initiation pertain to *chemical reactions* generating free radicals. Actually, many direct oxidation processes may involve free radicals as intermediates. The latter might be capable of reacting with O_2, forming radicals of the type $ROO\cdot$.

As far as commercial plastics are concerned, the thermal decomposition of hydroperoxides is considered to be of utmost importance. Hydroperoxides can be present in synthetic polymers in minute amounts, either incorporated chemically into the polymer (during processing at elevated temperatures) or admixed to the polymer as low molecular weight compounds (impurities). Many hydroperoxides decompose at relatively low temperatures according to reaction (7.17):

$$ROOH \rightarrow RO\cdot + \cdot OH \tag{7.17}$$

$\cdot OH$ radicals are very reactive. They react with many organic substances with encounter-controlled rate constants. Commonly they either abstract hydrogen atoms (in the case of aliphatic compounds) or undergo addition reactions (e.g. in the case of aromatic compounds). Regarding polymer degradation, H atom abstraction processes proceeding according to reaction (7.18) are most important:

$$RH + \cdot OH \rightarrow R\cdot + H_2O \tag{7.18}$$

It should be recalled here that, at sufficiently high oxygen concentrations, reaction (7.18) will be followed by

$$R\cdot + O_2 \rightarrow ROO\cdot \tag{7.19}$$

and

$$ROO\cdot + RH \rightarrow ROOH + R\cdot \tag{7.20}$$

Reactions (7.19) and (7.20) correspond to the propagation steps of autoxidation processes.

7.4.3 Influence of Metals

Frequently, trace quantities of metals catalyze oxidation processes in polymers. Typical results obtained with polypropylene in solution at 125°C are presented in Fig. 7.3. Metal impurities arise through contamination from reaction or storage vessels. In certain cases they are remainders of polymerization catalysts. It is generally accepted that metals are catalytically active in the ionized state. Commonly, multivalent

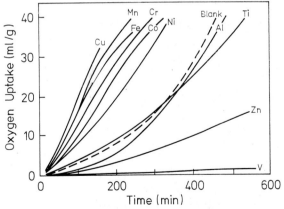

Fig. 7.3 The influence of metal ions on the oxidation of isotactic polypropylene in 1,2,4-trichlorobenzene solution (2.38 base mol/l) at 125°C [Catalyst]: 7.9×10^{-4} mol/l. (After *Osawa* and *Saito* [69])

metal ions operate as catalysts, for example, manganese ions catalyze the autoxidation of polyamide-6 [64] and iron ions stimulate the autoxidation of GR-S rubber [65]. As has been pointed out by Reich and Stivala [63], metal catalysts can affect the autoxidation process according to various mechanisms depending on experimental conditions, such as medium, type of metal salt, metal ion concentration, etc.

The main function of metal ions consists in inducing the decomposition of hydroperoxides by redox reactions, thus generating free radicals:

$$R-O-OH + Me^{n(+)} \to R-O\cdot + Me^{(n+1)(+)} + OH^{(-)} \tag{7.21}$$

$$R-O-OH + Me^{(n+1)(+)} \to R-O-O\cdot + Me^{n(+)} + H^{(+)} \tag{7.22}$$

Typical modes of action are presented in Scheme 7.4.

In the case of low molecular weight *unsaturated* hydrocarbons, evidence was obtained for the formation of complexes with certain metal ions resulting in an acceleration of the rate of oxygen uptake [66, 67].

Scheme 7.4   Modes of action of metal ions during the decomposition of hydroperoxides (adopted from *Reich* and *Stivala* [63])

Metal ions are strong oxidizing agents:

$$ROOH + (CH_3COO)_4Pb \to RO_2^\cdot + CH_3COOH + (CH_3COO)_3Pb$$

Metal ions are strong reducing agents:

$$ROOH + Fe^{2(+)} \to RO\cdot + Fe^{3(+)} + OH^{(-)}$$

Metal ions possess valence states of comparable stability:

$$ROOH + Co^{2(+)} \to RO\cdot + Co^{3(+)} + OH^{(-)}$$

$$ROOH + Co^{3(+)} \to ROO\cdot + Co^{2(+)} + H^{(+)}$$

A frequently encountered case pertains to oxidative processes at metal/polymer interfaces. In this case the enhancement of oxidation rates depends critically upon the rate of dissolution of metal ions, which, in turn, depends on the rate of diffusion of the ions into the polymer matrix. According to a mechanism elaborated for the system polyethylene/copper by *Allara* and *White* [68], the metal salt diffuses into the matrix and interacts with impurity carboxyl groups (PCOOH) at the polymer chains, thus releasing acid (RCOOH(m)) which can diffuse to the surface to form salt again (see Scheme 7.5).

The reclamation (recycling) of vulcanized rubber via metal ion-initiated autoxidation processes has recently been the subject of intense research. According to Yamashita et al. [132, 133], synthetic and natural rubber vulcanizates readily undergo oxidative degradation at room temperature if brought in contact (e.g. on soaking) with a solution containing a complex of phenylhydrazine and ferrous chloride. Scrap rubber, reportedly, can be reclaimed by this process. Elastomer waste of various vulcanizates subjected to this reclamation procedure yielded useful elastomers after revulcanization, e.g. synthetic isoprene rubber, acrylonitrile-butadiene-rubber, ethylene-propylene terpolymer (EPDM), butyl rubber, styrene-butadiene rubber etc.

Scheme 7.5   The dissolution of a copper salt into the matrix, designated by (m), of polyethylene (according to *Allara* and *White* [68])

$$(RCOO)_2Cu \rightleftharpoons (RCOO)_2Cu(m)$$

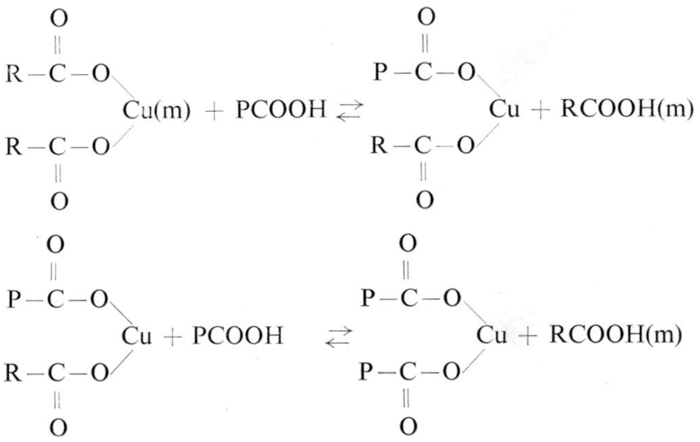

At present, however, processes operating without the use of chemical additives are applied to an increasing extent. In these processes, cured factory scrap or waste rubber products are converted — by mechanical means, even at ambient temperature — to a fine powder, which can be used as an additive for asphalt paving compositions or as a partial substitute for virgin rubber*).

7.4.4   Autoxidation of Polymers

In the last part of the section devoted to oxidative degradation processes some typical examples will be discussed. Saturated and unsaturated hydrocarbon polymers were selected because the oxidation of these polymers has been studied for a rather long time. The fact that, up to the present day, extensive activities in various laboratories are concerned with the autoxidation of hydrocarbon polymers demonstrates, that there are still unsolved problems. It must be emphasized that the situation concerning hydrocarbon polymers is symbolic of the situation existing with other polymers. The reason for this lack of understanding is that oxidative processes commonly proceed in heterogeneous systems (except at temperatures above the melting point). Oxygen diffusion limitations, the involvement of secondary reactions and/or selective action in partially crystalline polymers, usually cause significant deviations from the simple and idealized reaction scheme presented for autoxidation.

Fig. 7.4 (a) shows a typical exponential curve indicating the acceleration of the oxygen uptake during the reaction [70]. This curve, obtained with a low molecular weight hydrocarbon, is drastically changed if the reaction products are not continuously removed. This becomes evident from Fig. 7.4 (b), which depicts a sigmoidal curve, obtained with-

---

*) *R. A. Swor, L. W. Jenson* and *M. Budzol,* Rubber Chem. Technol, 53, 1215 (1980); T. C. P. Lee W. Millus, US Pat. 4,046,834 (1977) and literature cited therein.

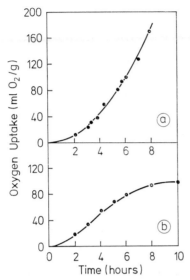

Fig. 7.4 Autoxidation of cetane at 140 °C, (a) with removal of secondary products, (b) without removal of secondary products. (After *Yurev, Pravednikov* and *Medvedev* [70])

out removal of secondary products. Sigmoidal curves are also quite typical for the dependence of oxygen uptake or the formation of oxidized products in polymers undergoing autoxidation, as demonstrated by the curve for the blank in Fig. 7.3. Fig. 7.5 shows that the induction period (in this case for the autoxidation of branched polyethylene at 140 °C) can be significantly prolonged by the presence of an inhibitor (e.g. activated carbon). The inhibitor interferes here with the propagation process.

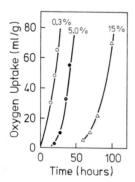

Fig. 7.5 Autoxidation of branched polyethylene in the presence of carbon black containing bound oxygen. (After *Winslow* [71])

Saturation values of the O_2-uptake, being approached after prolonged treatment, can be due to the consumption of oxygen in the sample or due to the inaccessibility of reactive sites in the polymer specimen. Fig. 7.6 shows how the ultimate oxygen uptake in polyethylene increases as crystallites, impervious to oxygen, pass through fusion/recrystallization stages with rising temperature. If crystallites are permeable to oxygen, as with poly-4-methylpentene-1, the oxidation patterns for amorphous and crystalline regions are the same [71].

Among the main products of autoxidation of saturated hydrocarbon polymers are hydroperoxides and their decomposition products. For polyethylene (PE) and polypropylene (PP), for example, formic acid, acetic acid and acetone were found as major volatile products [72, 73]. Furthermore, it was revealed that many adjacent hydroperoxide groups in the chains were produced in the oxidation of PE and PP [74]. In kinetic studies it turned out that intramolecular propagation reactions are rather im-

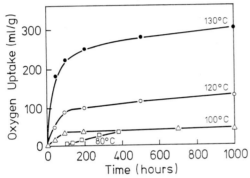

Fig. 7.6 Autoxidation of linear polyethylene ($\overline{M}_{w,o} > 10^6$) at various temperatures. (After *Winslow* [71])

portant [75, 76]. By electron spin resonance (ESR) measurements, peroxyl radicals could be identified in PP and poly-1-butene [77]. The lifetime of these radicals was estimated as 10^{-2} s. Owing to their short lifetime of probably less than 10^{-8} s [78], alkoxyl radicals could not be detected. It should be pointed out here that alkoxyl radicals decompose by β-scission:

$$\text{∿CH}_2\text{—CH—CH}_2\text{—CH}_2\text{∿} \longrightarrow \text{∿CH}_2\text{—C}{\overset{O}{\underset{H}{\diagup}}} + \text{·CH}_2\text{—CH}_2\text{∿} \qquad (7.23)$$

The occurrence of reaction (7.23) implies a diminution of the molecular weight due to main-chain rupture, with a consequential drastic deterioration in mechanical properties. Alkoxyl radicals are generated in termination reactions between pairs of peroxyl radicals

$$PO_2^{\cdot} + PO_2^{\cdot} \rightarrow 2\,PO^{\cdot} + O_2 \qquad (7.24)$$

Therefore, main-chain cleavage will be inhibited if peroxyl radicals are scavenged by antioxidants. Such additives, in rather low concentrations, can prevent a rapid deterioration of mechanical properties, whereas the formation of oxidation products is only slightly influenced, because the propagation process can comprise many kinetic steps before being terminated by reaction (7.24) or by another radical-radical reaction. Main-chain cleavage will be inhibited if the inhibitor interferes with the termination step.

Rate constants for propagation (k_p) and for termination (k_t) were determined for polyolefins with the aid of ESR measurements by *Chien* and *Wang* [77 (b)]. In order to control reaction conditions, autoxidation was initiated by thermal decomposition of benzoyl peroxide at definite temperatures between 71° and 105 °C. The temperature dependence of the rate constants for polypropylene is given by Equations (7.25) and (7.26):

$$k_p = 1.6 \times 10^7 \exp(-12080/RT) \qquad (7.25)$$

$$k_t = 1.11 \times 10^{13} \exp(-11600/RT) \qquad (7.26)$$

(Rate constants in (l/mol s) and activation energies in (cal/mol degree). The temperature dependence of autoxidation processes in polyolefins under natural conditions has been assessed recently by *Demisov* [79].

Table 7.6   Autoxidation of polyisoprene. Principal volatile products resulting from chain scission (adopted from *Morand* [83])

Oxidation products	Formulae
methacrolein	$\begin{array}{c} CH_2 \\ \parallel \\ H_3C-C-C-H \\ \parallel \\ O \end{array}$
methylvinyl ketone	$\begin{array}{c} H_3C-C-CH=CH_2 \\ \parallel \\ O \end{array}$
levulinaldehyde	$\begin{array}{c} H \\ \vert \\ H_3C-C-CH_2-CH_2-C=O \\ \parallel \\ O \end{array}$
4-hydroxy-2-butanone	$\begin{array}{c} H_3C-C-CH_2-CH_2OH \\ \parallel \\ O \end{array}$
4-hydroxy-4-methyl-5-hexenal	$\begin{array}{c} CH_3 H \\ \vert \vert \\ CH_2=CH-C-CH_2-CH_2-C=O \\ \vert \\ OH \end{array}$
4-methyl-4-vinyl-butyrolactone	$\begin{array}{c} \ CH_2{-}CH_2 \\ H_3C\vert\vert \\ \diagdown C\diagup C\diagdown \\ \vert\ \diagdown O \diagup\ O \\ H_2C=CH \end{array}$
5-hydroxy-6-methyl-6-heptene-2-one	$\begin{array}{c} CH_3 CH_3 \\ \vert \\ CH_2=C-CH-CH_2-CH_2-C-CH_3 \\ \vert \parallel \\ OH O \end{array}$

For many years, autoxidation processes in unsaturated hydrocarbon polymers have attracted the interest of various researchers. Polyisoprene, being the main constituent of natural rubber, was given special attention. The autoxidation of this polymer has been

Scheme 7.6 Autoxidation of polyisoprene by a free radical chain mechanism [83–85]

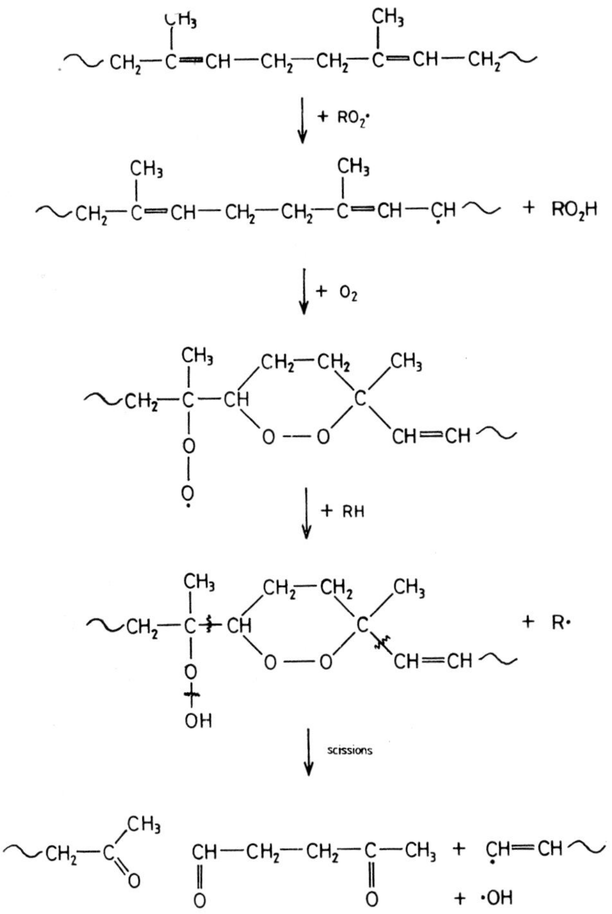

$$\begin{aligned}
&\overset{CH_3}{\underset{|}{}}\quad\overset{CH_3}{\underset{|}{}}\\
&\sim CH_2-C\!=\!CH-CH_2-CH_2-C\!=\!CH-CH_2\sim
\end{aligned}$$

$\downarrow$ + RO$_2$·

$$\sim CH_2-\underset{\overset{|}{CH_3}}{C}\!=\!CH-CH_2-CH_2-\underset{\overset{|}{CH_3}}{C}\!=\!CH-\overset{\cdot}{C}H\sim \quad + \ RO_2H$$

$\downarrow$ + O$_2$

$$\sim CH_2-\underset{\overset{O}{\underset{\overset{|}{\overset{O}{\cdot}}}{|}}}{\overset{|}{\underset{CH_3}{C}}}-CH\underset{O-O}{\overset{CH_2-CH_2}{\diagup\diagdown}}C\underset{CH=CH\sim}{\overset{CH_3}{\diagup\diagdown}}$$

$\downarrow$ + RH

$$\sim CH_2-\underset{\overset{O}{\underset{\overset{|}{OH}}{+}}}{\overset{|}{\underset{CH_3}{C}}}\!+\!CH\underset{O-O}{\overset{CH_2-CH_2}{\diagup\diagdown}}C\underset{CH=CH\sim}{\overset{CH_3}{\diagup\diagdown}} \quad + \ R·$$

$\downarrow$ scissions

$$\sim CH_2-C\underset{O}{\overset{CH_3}{\diagup\diagdown}} \quad \underset{\overset{\|}{O}}{CH}-CH_2-CH_2-\underset{\overset{\|}{O}}{C}-CH_3 \ + \ \overset{\cdot}{C}H\!=\!CH\sim$$
$$+ \ ·OH$$

studied systematically up to a temperature of 340 °C [80–85]. The principal volatile products liberated on heating purified synthetic polyisoprene at 100 °C in air are listed in Table 7.6 [83]. It appears that these products result from chain scission in the polymer. At higher temperatures secondary decompositions become important. The most abundant products formed at 268 °C to 340 °C are [84]: methyl ethyl ketone, methyl vinyl ketone and butyraldehyde with smaller amounts of acetaldehyde, acetone, methacrolein, propionaldehyde, acrolein and formaldehyde. At present it is generally accepted that the autoxidation of polyisoprene occurs according to free radical chain mechanisms.

A typical mechanism is presented in Scheme 7.6.

7.5   Ionic Degradation

Ionic reactions also play a significant role in chemical degradation. Remarkable progress has been made recently concerning both evidence for and elucidation of ionic processes. Therefore, it is appropriate to devote this section to a discussion of a few typical cases. One of these is the acidic hydrolysis already discussed in Section 7.2. Further examples, selected for this discussion, are (a) the alkaline degradation of polysaccharides, (b) the acidic degradation of polyaldehydes and polyacetals, (c) the cationic degradation of poly(propylene sulfide) and polyesters.

(a) The alkaline degradation of polysaccharides has been subject of various studies during the last decades [86—96]. The degradation proceeds as an "unzipping" reaction, also called "endwise degradation". The reaction, initiated at *reducing terminal base units*, involves the liberation of end units which are further modified to saccharinic acids, e.g. D-gluco-*iso*-saccharinic acid. "Unzipping" continues until halted by a termination reaction, e.g. by the formation of alkali-stable *meta*-saccharinic acid (MSA) residues. Alternatively whole macromolecules are decomposed. A mechanism proposed by *Lai* and *Sarkanen* [93] (corroborated recently by *Ziderman* and *Bel-Ayche* [95]), is presented in Scheme 7.7.

Scheme 7.7   Mechanism of unzipping of cellulose, catalyzed by hydroxyl ions [93, 95]. G: glucose unit; MSA: *meta*-saccharinic acid

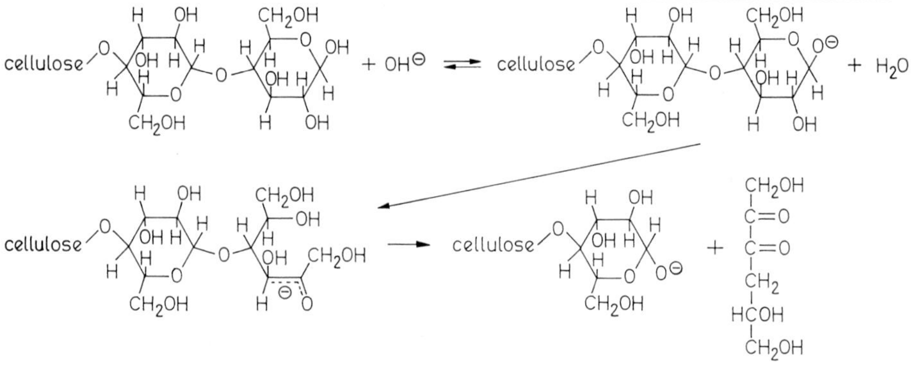

Initiation

$$H-G-O\{G-O\}_{\!n}G-OH + OH^{\ominus} \rightleftharpoons H-G-O\{G-O\}_{\!n}G-O^{\ominus} + H_2O$$

Propagation

$$H-G-O\{G-O\}_{\!n}G-O^{\ominus} \longrightarrow H-G-O\{G-O\}_{\!n-1}G-O^{\ominus} + products$$

Termination

$$H-G-O^{\ominus} \longrightarrow products$$

$$H-G-O\{G-O\}_{\!m}G-O^{\ominus} \longrightarrow H-G-O\{G-O\}_{\!m}MSA$$

$$m \leq n$$

The number of unzipped glucose units correspond to $40-70$ in the case of cellulose (heterogeneous reaction) or to several hundred in the case of amylose (treated in solution) [92]. In the latter case unzipping is essentially limited by branch points at carbons in 2 or 3 positions (compare Section 6.3.1.1).

Scheme 7.8   Acidic degradation of polyacetaldehyde

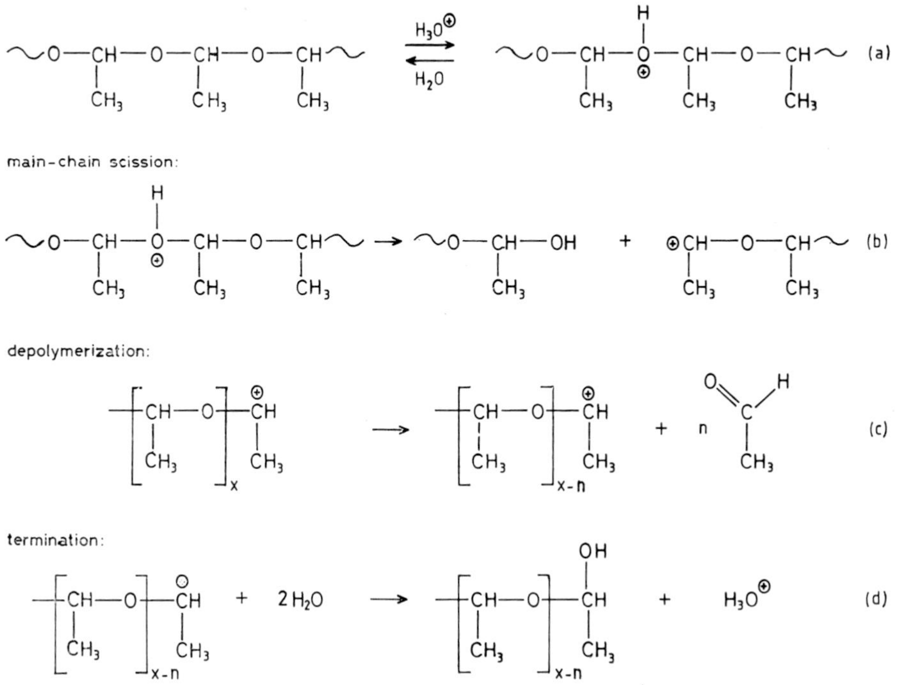

(b) In the acidic degradation of polyaldehydes with the general structure

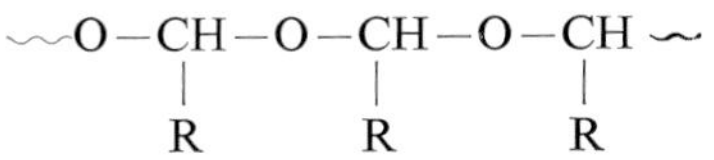

$$\sim O-CH-O-CH-O-CH\sim$$
$$\quad\quad\ | \quad\quad\quad | \quad\quad\quad |$$
$$\quad\quad R \quad\quad\ \ R \quad\quad\ \ R$$

two successive steps have been distinguished. The initial step causes main-chain scission. In the following step depolymerization occurs. A typical polymer exhibiting this behavior is polyacetaldehyde ($R = CH_3$), whose degradation in dilute ethyl acetate solution in the presence of acetic acid was investigated by Delzenne and Smets, about 25 years ago [97]. The mechanism is illustrated in Scheme 7.8. Activation energies for the two processes were determined as about 84 kJ/mol (main-chain rupture) and about 10 kJ/mol (depolymerization). An analogous behavior was found for polyformaldehyde [98] and poly-1,3-dioxolane [99]. In the latter case acidolysis of the polymer leads to the formation of dioxolane:

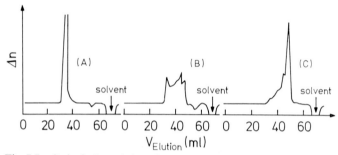

$$(7.27)$$

Upon treating polydioxolane with picric acid (0.1 to 1.0 weight%) between 140° and 200°C, the number of dioxolane molecules, formed per chain cleavage, was found to be about 10 [99].

Fig. 7.7   Cationic degradation of poly(propylene sulfide) in CH_2Cl_2 at 20°C induced by triethyloxoniumtetrafluoroborate (10^{-2} mol/l). [Polymer]: 19.6 g l^{-1}. (A): Directly after initiation, (B): 30 min and (C) 30 h after initiation. (After *Simonds* and *Goethals* [100])

(c) Polythiopropylene, or poly(propylene sulfide), degrades rapidly if treated with tri-ethyloxonium tetrafluoroborate (TEFB) [100]. The decrease in MW can be inferred from the gel permeation chromatograms shown in Fig. 7.7. The reaction was carried out at 20°C in CH_2Cl_2. In the first stage the average molecular weight decreased. Ulti-mately a mixture of low molecular weight compounds, comprising cyclic tetramers and pentamers, several isomers of 3,6-dimethyl-1,2,5-trithiepane, and propene, was formed:

The proposed mechanism depicted in Scheme 7.9 is based on the assumption that epi-sulfonium ions exist as intermediates [100 (b)]. Similar results as with poly(propylene sulfide) have been obtained with the following polymers (polythiiranes) [100 (c)]:

Scheme 7.9 Cationic degradation of poly(propylene sulfide) [100]

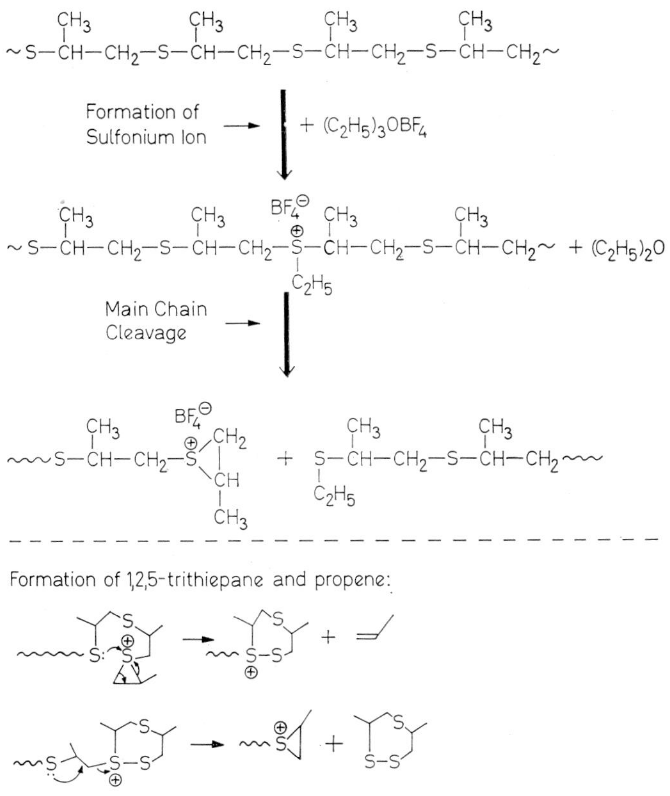

$$\sim S-\underset{\underset{CH_3}{|}}{CH}-CH_2-S-\underset{\underset{CH_3}{|}}{CH}-CH_2-S-\underset{\underset{CH_3}{|}}{CH}-CH_2-S-\underset{\underset{CH_3}{|}}{CH}-CH_2\sim$$

Formation of Sulfonium Ion $\longrightarrow$ $+ (C_2H_5)_3OBF_4$

$$\sim S-\underset{\underset{CH_3}{|}}{CH}-CH_2-S-\underset{\underset{CH_3}{|}}{CH}-CH_2-\overset{BF_4^\ominus}{\underset{\underset{C_2H_5}{|}}{\overset{\oplus}{S}}}-\underset{\underset{CH_3}{|}}{CH}-CH_2-S-\underset{\underset{CH_3}{|}}{CH}-CH_2\sim \;+ (C_2H_5)_2O$$

Main Chain Cleavage $\longrightarrow$

$$\sim\!\sim\!\sim S-\underset{\underset{CH_3}{|}}{CH}-CH_2-\overset{BF_4^\ominus}{\overset{\oplus}{S}}\underset{\underset{CH_3}{\overset{|}{CH}}}{\overset{CH_2}{\diagdown}} \;+\; S-\underset{\underset{C_2H_5}{|}}{CH}-CH_2-S-\underset{\underset{CH_3}{|}}{CH}-CH_2\sim\!\sim\!\sim$$

(with CH_3 on second structure)

- -

Formation of 1,2,5-trithiepane and propene:

Recent investigations [101] demonstrated that polyesters are readily degraded if subjected to the attack of carbocations (carbenium ions). In the reported experiment, p-chlorobenzylium cations, generated by the reaction

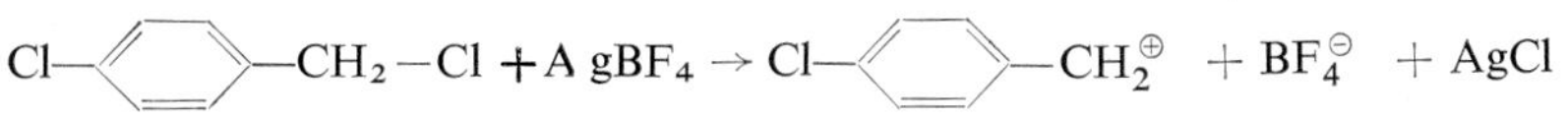

$$Cl\!-\!\!\left\langle\right\rangle\!\!-\!CH_2-Cl + A\,gBF_4 \rightarrow Cl\!-\!\!\left\langle\right\rangle\!\!-\!CH_2^\oplus + BF_4^\ominus + AgCl \qquad (7.28)$$

were reacted with poly(p-xylylene succinate)

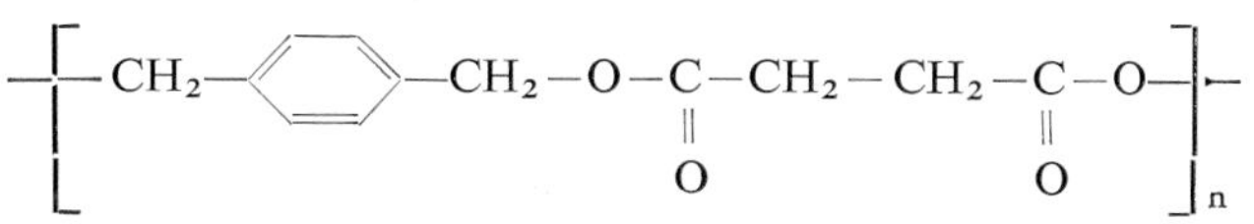

$$-\!\!\left[\!CH_2\!-\!\!\left\langle\right\rangle\!\!-\!CH_2-O-\underset{\underset{O}{\|}}{C}-CH_2-CH_2-\underset{\underset{O}{\|}}{C}-O\!\right]_{n}\!-$$

in methylene dichloride at 293 K. As evidenced by GPC the MW decreased indicating cleavage of ester linkages. It was concluded that carbenium ions are far more effective in cleaving ester linkages than protons. Although the mechanism could not be elucidated so far, three possible reaction routes were suggested [101]

$$R^\oplus + \begin{matrix} O=C-R_2 \\ | \\ O-R_1 \end{matrix} \begin{cases} \longrightarrow R-O-R_1 + O\overset{\oplus}{=\!\!=}C-R_2 \\ \longrightarrow R-O-C-R_2 + R_1^\oplus \\ \qquad\qquad \| \\ \qquad\qquad O \\ \longrightarrow O=C-R_2 + R_1^\oplus \\ \qquad\quad | \\ \qquad\quad R-O \end{cases} \qquad (7.29)$$

7.6   Selective Degradation

Selective action of chemical agents on polymers refers to two aspects: one dealing with the morphology of rigid polymers, the other with the specific attack of certain sites of the macromolecules.

Selective oxidation of partially crystalline polymers, such as polyethylene and isotactic polypropylene, proved to be extremely useful for the quantitative analysis of the texture of these polymers.

Many polymers crystallize by a folded-chain mechanism as depicted in Fig. 7.8. The amorphous phase can consist of macromolecules physically adsorbed on surface sites of single crystals consisting of regularly folded chains, and/or of loose loops which are formed if adjacent re-entry is prevented during the crystallization process [102–105]. Crystallites can be connected by tie-molecules, i.e. molecules with portions in regular folded regions of two (or more) different crystallites (not indicated in Fig. 7.8, but see Fig. 3.7).

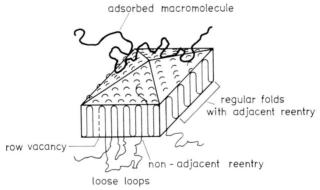

Fig. 7.8   Schematic illustration of a polymer single crystal in contact with amorphous phase structures. (According to *Hoffman* and *Davis* [102])

About 30 years ago, chemical etching techniques were first used successfully in the analysis of cellulose, by acid hydrolysis [111]. That way crystallites could be segregated from amorphous portions of the polymer. Later *Palmer* and *Cobbold* discovered [106] that fuming nitric acid preferentially attacks the amorphous regions (between the lamellar crystals) in polyethylene. As the reaction is not confined to the specimen's

Table 7.7 Selective degradation of partially crystalline polymers

polymer	structural repeating unit	method of degradation	Ref.
polyethylene	$-CH_2-CH_2-$	nitric acid ozone	106, 112, 113 107
isotactic polypropylene	$\begin{array}{c} CH_3 \\ \| \\ -CH_2-CH- \end{array}$	nitric acid	114
cellulose		acidic hydrolysis	111, 115
polyethyleneterephthalate	$-O-\overset{\|}{\underset{O}{C}}-\!\!\!\!\bigcirc\!\!\!\!-\overset{\|}{\underset{O}{C}}-O-CH_2-CH_2-$	monoethylamine	116—118
polyoxymethylene	$-O-CH_2-$		102, 119
poly(tetramethyl-p-silphenylene siloxane)	$\begin{array}{c} CH_3 \quad\quad CH_3 \\ \| \quad\quad\quad \| \\ -Si-\!\!\bigcirc\!\!-Si-O- \\ \| \quad\quad\quad \| \\ CH_3 \quad\quad CH_3 \end{array}$	acidic hydrolysis (HF)	110

surface, lamellar crystals can be separated for further study from relatively massive specimens by fuming nitric acid treatment. This method has been widely utilized for the study of certain partially crystalline polymers. Some typical polymers are listed in Table 7.7, which shows, furthermore, that various selective degradation methods have been used to investigate the nature of the fold structure, and particularly the nature of the fold surface of single crystals.

Upon prolonged acidic treatment, the chemical attack is usually not restricted to the bulk amorphous fraction, but regular fold sites are also attacked. Thus, it is possible to determine single traverse lengths of folded structures. Typical results are presented in Fig. 7.9, showing gel permeation chromatograms of polyethylene single crystals after fuming nitric acid treatment [104]. In this case single- and double-traverse peaks can be clearly distinguished.

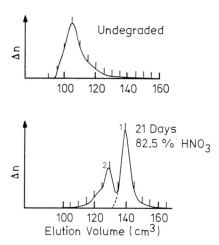

Fig. 7.9 Selective degradation of polyethylene. Gel chromatograms of undegraded and nitric acid degraded monolayer single crystals. Single-traverse and double traverse peaks are marked by 1 and 2, respectively. (According to *Keller* et al. [104])

The change of crystallinity with progressive etching is illustrated in Fig. 7.10, where results obtained with poly(tetramethyl-p-silphenylene siloxane)*) are presented [110]. Treatment of the polymer with hydrofluoric acid (HF) caused the hydrolysis of Si—O bonds. From Fig. 7.10 it is seen that the crystallinity of the residual polymer increases to ca. 95 % (while simultaneously the average molecular weight decreases). This corresponds to the removal of two or more layers of base units from each crystal surface [110]. Above 50% weight loss, the crystallinity decreases again, indicating that HF can now penetrate the specimen and attack the crystalline core.

Indications for non-random main-chain cleavage in a homogeneous system were found recently during the acidic hydrolysis of dextran in aqueous solution [120]. The polymer was treated at 80 °C with 0.12N sulfuric acid at polymer concentrations varying between

---

*) The structure of the repeating unit is given in Table 7.7.

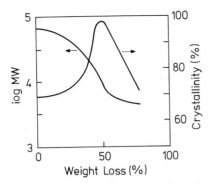

Fig. 7.10   Selective degradation of poly(tetramethyl-p-silphenylene siloxane) by 48% hydrofluoric acid at 20° to 50 °C. Average molecular weight (left ordinate) and degree of crystallinity (right ordinate) vs. conversion as indicated by weight loss. (After *Okui* and *Magill* [110])

0.25 and 2%. Reaction rate constans were proportional to $(MW)^{2/3}$. MWD curves (see Fig. 7.11), pertaining to early stages of the hydrolysis, possess a shoulder in the low MW range, indicating that small fragments are formed with a much higher probability than could be expected from a random attack of the base units. Thus, it was concluded that bonds located near the ends of the chain are more reactive than those close to the center.

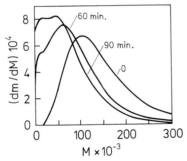

Fig. 7.11   Selective degradation of dextran $(\overline{M}_{n,0} = \overline{M}_{w,0}/1.29 = 1.17 \times 10^5)$ in aqueous solution (0.12 N sulfuric acid) at 80 °C. Polymer concentration: 1%. Differential MWD of undegradaded and degraded dextran. The duration of treatment is indicated in the graph. (After *Basedow, Ebert, Ederer* [120])

Significant selectivity effects were detected during the study of chemical transformations of three-dimensionally crosslinked systems forming *gels* in appropriate swelling agents. If ethylene glycol methacrylate gels, e.g., are subjected to alkaline hydrolysis, two stages of saponification can be distinguished [109]. In the initial stage, only ester linkages between the polymer backbone and side chains with non-capped hydroxyl groups are ruptured. Crosslink ester linkages are attacked afterwards at a much lower rate. The relevant reactions are illustrated in Scheme 7.10.

Scheme 7.10   Alkaline hydrolysis of ethylene glycol methacrylate gels
(according to Ševčik et al. [109])

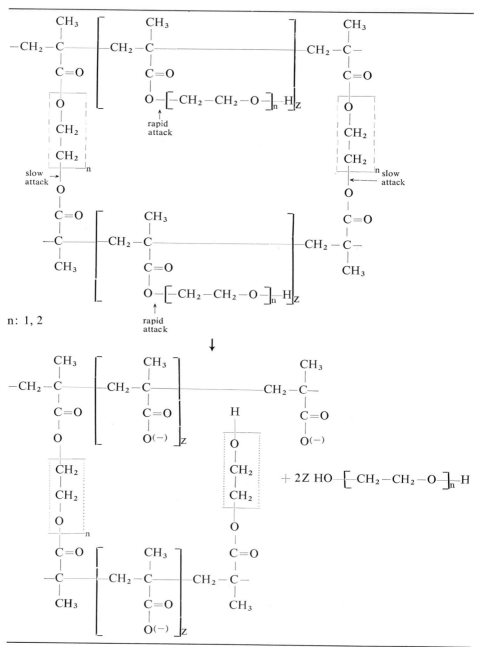

n: 1, 2

This selective action was assessed recently by investigating the dependence of the cross-linking density of glycol methacrylate gels on the degree of saponification. The cross-linking density remained constant during the initial stage and decreased only during the later stages of the process. A contrary behaviour was exhibited by low molecular weight model compounds: in ethylene glycol diacetate, for example, the first ester linkage was cleaved with a rate constant twice as high as that for the cleavage of the second ester linkage [123]. The polymer effect is assumed to be due to steric hindrance and in addition to differences in hydrophilicity between low and high molecular weight diester compounds [121, 122].

Concluding this section it should be emphasized that the examples discussed above are symbolic of a greater number of known cases.

Frequently, specific reactivity originates from the existence of functional groups at specific sites of the macromolecules. A prominent example for the latter case is endwise degradation (unzipping), which has been discussed in the preceding Section 7.5 in connection with the depolymerization of polysaccharides in alkaline media. In this case the polymers possess terminal functional groups capable of undergoing reactions with hydroxyl ions.

7.7   Reactions with Air Pollutants

In the course of investigations of the detrimental effects of air pollutants, polymers have attracted some interest since many plastics are subjected to prolonged or perpetual outdoor exposure. The investigations concentrated on the behavior of polymers in atmospheres containing nitrogen dioxide (NO_2) and sulfur dioxide (SO_2). Systems containing simultaneously NO_2, SO_2, and O_2 were also studied, as was the influence of UV light. Many relevant studies have been carried out by Jellinek et al. and were reviewed recently [124].

Upon treating various polymers with either NO_2 or SO_2 in the dark at $25°$ or $35°C$, saturated polymers were found to be rather unreactive on the basis of intrinsic viscosity measurements [125]. Among the saturated polymers studied were polyethylene, polypropylene, polyvinylchloride, and polyvinylpyrrolidone. On the other hand, polyamide-6,6 and linear polyurethane of the structure

$$\{(CH_2)_4 O-CO-NH-(CH_2)_6 CO-O\}_n ,$$

when subjected to the same treatment, were affected [128].

The attack of the latter polymer resulted in simultaneous scission and crosslinking. Gel formation indicated predominant crosslinking [126].

Infrared studies revealed that polyethylene reacts with NO_2 at elevated temperatures [127]. Conversions observed at $25°C$ are probably due to reactions with (impurity) olefinic double bonds which react readily with NO_2 (see below). Polystyrene was reported [124] to react with NO_2. The rate constant of the reaction

$$-CH-CH_2- + NO_2 \rightarrow -\overset{\bullet}{C}-CH_2- + HNO_2 \qquad (7.30)$$
$$\quad\; |\qquad\qquad\qquad\qquad\quad\; |$$
$$\;\; Ph\qquad\qquad\qquad\qquad\quad Ph$$

depends on temperature and partial pressure of $NO_2 (P_{NO_2})$ according to Eq. (7.31):

$$k_{PSt+NO_2} = P_{NO_2}\, 5.52 \times 10^4 \exp\left(-\frac{16\,200}{RT}\right)\, (h^{-1}) \qquad (7.31)$$

The mechanism postulated for the reaction of NO_2 with polystyrene and polyethylene is shown in Scheme 7.11.

Scheme 7.11   Degradation of polyethylene and polystyrene by NO_2

$PH + NO_2 \longrightarrow P\cdot + HNO_2$	hydrogen abstraction
$P\cdot + NO_2 \; \begin{array}{l}\longrightarrow PNO_2 \\ \longrightarrow P{-}O{-}N{=}O\end{array}$	addition of nitro groups formation of nitrite ester groups
$PONO \longrightarrow PO\cdot + NO$	rupture of $O{-}N$ bond
$PO\cdot + NO_2 \longrightarrow PONO_2$	formation of nitrate ester groups
$PO\cdot \longrightarrow F_1 + F_2^{\cdot}$	main-chain cleavage (β-scission)

$$\left(PO\cdot \longrightarrow \quad \sim\!\!\overset{\displaystyle\overset{H}{|}}{C}{=}O + \cdot CH_2\!\!\sim\right)$$

$PO\cdot + PH \longrightarrow POH + P\cdot$	formation of hydroxyl groups

Unsaturated hydrocarbon polymers are readily attacked by both SO_2 and NO_2: butyl rubber undergoes predominantly main-chain scission whereas polyisoprene crosslinks NO_2 and SO_2 react by addition to double bonds:

$$\begin{array}{c}\rangle C{=}C\langle + NO_2 \longrightarrow -\underset{\cdot}{C}-\underset{\underset{NO_2}{|}}{C}- \end{array} \qquad (7.32)$$

Most of the reactions of SO_2 with polymers were reportedly [124] strongly enhanced by simultaneous irradiation with UV light, which is absorbed by SO_2:

$$SO_2 + hv \longrightarrow {}^1SO_2^* \longrightarrow {}^3SO_2^* \qquad (7.33)$$

Triplet excited sulfur dioxide (${}^3SO_2^*$) is assumed to be capable of abstracting hydrogen:

$$ {}^3SO_2^* + PH \longrightarrow P\cdot + H\dot{S}O_2 \qquad (7.34)$$

Macroradicals generated by reaction (7.34) can undergo various reactions, e.g.

$$P\cdot + SO_2 \longrightarrow PSO_2^{\cdot} \qquad (7.35)$$

or, in the presence of O_2:

$$P\cdot + O_2 \longrightarrow PO_2^{\cdot} \qquad (7.36)$$

7.8   Solvent Stability

Problems concerning the stability of plastics towards solvents are related only partly to chemical degradation. As was outlined in Section 7.1, strong physical interaction frequently leads to a deterioration of polymeric specimens. Physical interactions can cause swelling and dissolution of the polymer.

Macromolecular systems consisting of spatial networks are insoluble but often capable of swelling. This behavior pertains to the large group of thermosetting polymers, where the degree of swelling depends not only on the extent of physical interaction between solvent and polymer, but also on the crosslinking density (number of crosslinks per repeating unit).

Most linear polymers (thermoplastics) are soluble in several solvents. Usually, a swelling stage precedes molecular dispersion when a linear polymer is brought in contact with a solvent.

Macromolecules capable of forming crystallites are rather resistant against physical interaction with solvents. Dissolution is impeded owing to strong intermolecular interactions between macromolecules and, unless these interactions are overcome by heating, the polymers are insoluble. Typical polymers of this sort, that dissolve at elevated temperatures only, are polyethylene and ordered polypropylene. Polytetrafluoroethylene is in soluble in all solvents, although it is not chemically crosslinked. Polyvinylalcohol becomes soluble only after hydrogen bonds have been destroyed beforehand by heating at 100 °C.

In many cases polymers do not remain intact during dissolution, but are chemically attacked by the low molecular weight liquid, causing decomposition into fragments. Quite important examples of such chemical attack are acidic and alkaline degradation, which have been dealt with in some detail in Section 7.2, in connection with the discussion of solvolysis phenomena.

In practical applications it is not necessary, in general, to distinguish between physical and chemical deterioration. The user's interest merely concerns the question whether or not a polymer is attacked by a low molecular weight liquid. Table 7.8 shows the solvent stability of typical polymers.

Table 7.8 The resistance of typical polymers to solvents at ambient temperature, based on immersion tests (adopted from [129] and [130])

Solvent \ Polymer	PE	PSt	PA-6	PVC	PTFE	EP-R	UP-R	SIL-R
acetone	L	U	S	U	S	U	L	U
methanol	S	S	S	S	S	S	S	S
saturated hydrocarbons	U	L	S	S	S	S	S	U
benzene	U	L	S	U	S	S	S	U
carbon disulfide	U			U	S	L	L	
carbon tetrachloride	U	U	S	U	S	L	S	U

Stability Code: S: satisfactory resistance, L: limited resistance, U: unsatisfactory resistance

Polymer Code: *PE*: polyethylene, *PSt*: polystyrene, *PA-6*: polyamide-6, *PVC*: polyvinylchloride, unplasticized, *PTFE*: polytetrafluoroethylene, *EP-R*: epoxy resin, *UP-R*: unsaturated polyester resin, *SIL-R*: silicon resin

Concerted actions can play an important role with respect to solvent stability of polymers, as can be seen from two examples: (a) mixtures of two or more organic liquids (mixed solvents) can develop solvent power towards a certain polymer. Separately applied, those liquids remain inactive. (b) One component of a mixture of liquids can attack a polymer chemically, while another acts as solvent for the decomposition product(s), or vice versa, the polymer is swollen by one liquid, which leads to a greater accessibility of reactive sites in the polymer for the other liquid.

In many cases simple immersion tests are sufficient for the characterization of polymers with respect to solvent resistance. For a variety of applications more specific characterizations are required, which take into account parameters such as temperature and duration of exposure. Sophisticated methods have, therefore, been developed, e.g. tests for chemical resistance under mechanical stress [131], in addition to the standard test methods used in various countries [27].

References to Chapter 7

[1] B. Doležel, "Die Beständigkeit von Kunststoffen und Gummi", C. M. von Meysenburg (ed.), Hanser, München (1978).
[2] H. H. G. Jellinek (ed.), "Aspects of Degradation and Stabilization of Polymers", Elsevier Amsterdam (1978).
[3] R. K. Eby, "Durability of Macromolecular Materials", ACS Symposium Series 95, American Chemical Society, Washington D.C. (1979).
[4] D. L. Allara and W. L. Hawkins (eds.), "Stabilization and Degradation of Polymers", Advances in Chemistry Series 169, American Chemical Society, Washington (1978).
[5] I. Mellan, "Corrosion Resistant Materials Handbook", Noyes Data Corporation, Park Ridge, N.J. (1976).
[6] E. M. Fettes, "Chemical Reactions of Polymers", Interscience-Wiley, New York (1964).
[7] N. Grassie, "Chemistry of High Polymer Degradation Processes", Butterworth, London (1956).
[8] H. H. G. Jellinek, "Degradation of Vinyl Polymers", Academic Press, New York (1955).
[9] H. A. Stuart, "Alterung und Korrosion von Kunststoffen", Verlag Chemie, Weinheim (1967).
[10] K. Thinius, "Stabilisierung und Alterung von Plastwerkstoffen", Verlag Chemie, Weinheim (1971).
[11] W. L. Hawkins (ed.), "Polymer Stabilization", Wiley-Interscience, New York (1972).
[12] L. Reich and S. S. Stivala, "Elements of Polymer Degradation", McGraw-Hill, New York (1971).
[13] G. Scott, "Atmospheric Oxidation and Antioxidants", Elsevier, New York (1965).
[14] S. H. Pinner (ed.), "Weathering and Degradation of Plastics", Gordon and Breach, London (1966).
[15] G. Schreyer, "Konstruieren mit Kunststoffen", Teil 2, Hanser, München (1972).
[16] J. C. West, Anal. Chem. 47, 1708 (1975).
[17] R. H. Greeley, J. Chrom. 88, 229 (1974).
[18] D. J. Mosselman and J. de Wit, Proc. IUPAC Symp. Polym. Chem., Dublin (1977), paper II-10.
[19] O. Mlejnik and L. Crečková, J. Chrom. 94, 135 (1974).
[20] G. G. Esposito and M. H. Swann, Anal. Chem. 34, 1048 (1962).
[21] D. F. Percival, Anal. Chem. 35, 236 (1963).

[22] *S. Mori, M. Furusawa* and *T. Takeuchi*, Anal. Chem. 42, 138 (1970).

[23] *J. L. Mulder*, Anal. Chim. Act. 38, 563 (1967).

[24] *W. Mazurek* and *G. C. Smith*, J. Oil Colour Chem. Ass. 57, 179 (1974).

[25] *M. D. Stevens*, "Characterization and Analysis of Polymers by Gas Chromatography", Dekker, New York (1969).

[26] *L. R. Whitlock* and *S. Siggia*, Sep. and Pur. Meth. 3, 299 (1974).

[27] DIN 53476; GOST 12020-66; ISO TC 612537 (1975); ASTM D 543-43 (1943).

[28] *J. Ehrbar* and *C.-M. Meysenburg*, Z. f. Werkstofftechnik 7, 429 (1976).

[29] *F. Micheel*, "Chemie der Zucker und Polysaccharide", Akademische Verlagsgesellschaft, Leipzig (1956).

[30] *H. Jalke*, "Entschlichtungsmittel" in *A. Chwala* and *V. Anger* (eds.), "Handbuch der Textilhilfsmittel", Verlag Chemie, Weinheim (1977).

[31] *J. Brandrup*, "Kunststoffe, Verwertung von Abfällen" in "Ullmanns Encyklopädie der technischen Chemie", 4th ed. Vol. 15, p. 411, Verlag Chemie Weinheim (1978); Kunststoffe 65, 881 (1975).

[32] *N. Calderon, J. P. Lawrence* and *W. A. Ofstead*, "Olefin Metathesis", Adv. Organometallic Chem. 17, 449 (1979).

[33] *R. H. Grubbs*, Progr. Inorg. Chem. 24, 1 (1978).

[34] *Th. J. Katz*, Adv. Organometallic Chem. 16, 283 (1977).

[35] (a) *G. Dall'Asta*, Makromol. Chem. 154, 1 (1972); Rubber Chem. Technol. 47, 511 (1974);
    (b) *K. J. Ivin, D. T. Laverty, J. H. O'Donnel, J. J. Rooney* and *C. D. Stewart*, Makromol. Chem. 180, 1975 and 1989 (1979);
    (c) *F. W. Küpper*, Angew. Makromol. Chem. 80, 207 (1979);
    (d) *V. W. Motz* and *M. F. Farona*, Inorg. Chem. 16, 2545 (1977).

[36] *Yu. V. Korshak, Al.Al. Berlin, E. R. Badamshina, G. I. Timofeeva* and *G. I. Pavlova*, Doklad. Akad. Nauk SSSR 248, 372 (1979).

[37] *W. Ast* and *K. Hummel*, Naturwissensch. 57, 545 (1970), Makromol. Chem. 166, 39 (1973).

[38] *W. Ast, C. Zott* and *R. Kerber*, Makromol. Chem. 180, 315 (1979).

[39] *F. Stelzer, K. Hummel* and *R. Thummer*, Prog. Colloid Polym. Sci. 66, 411 (1979).

[40] *E. Thorn-Csányi* and *H. Perner*, Makromol. Chem. 180, 919 (1979).

[41] *N. J. Pakuso, A. R. Gantmacher* and *B. A. Dolgoplosk*, Vysokomol. Soedin. Ser. B 20, 805 (1978).

[42] *K. Hummel* and *G. Raithofer*, Angew. Makromol. Chem. 50, 183 (1976).

[43] *J. Voigt*, "Die Stabilisierung der Kunststoffe gegen Licht und Wärme", Springer, Berlin (1966).

[44] *S. W. Benson*, Adv. Chem. Series 77, 74 (1968).

[45] *R. Criegee*, "Peroxide Mechanisms", Interscience, New York (1961).

[46] *R. W. Murray*, "Prevention of Degradation by Ozone", in *W.L. Hawkins* (ed.), "Polymer Stabilization", Wiley/Interscience, New York (1972).

[47] *J. R. Durland* and *H. Adkins*, J. Am. Chem. Soc. 61, 429 (1939).

[48] *C. C. Schubert* and *R. N. Pease*, J. Am. Chem. Soc. 78, 2044 and 5553 (1958).

[49] *L. Long* and *L. F. Fieser*, J. Am. Chem. Soc. 62, 2670 (1940).

[50] *M. C. Whiting, A. J. N. Bolt* and *J. H. Parish*, Adv. Chem. Series 77, 4 (1968).

[51] *G. A. Hamilton, B. S. Ribner* and *T. M. Hellman*, Adv. Chem. Series 77, 15 (1968).

[52] *M. I. Abdullin, R. F. Gataullin, K. S. Minsker, A. A. Kefeli, S. D. Razumorskii* and *G. Ye. Zaikov*, Vysokomol. soyed. A 13, 1143 (1977).

[53] *J. Suzuki, K. Hukushima* and *S. Suzuki*, Environm. Sci. Techn. 12, 1180 (1978).

[54] *J. Suzuki*, J. Appl. Polym. Sci. 20, 93 (1976).

[55] *C. C. Price* and *A. L. Tumolo*, J. Am. Chem. Soc. 86, 4691 (1964); J. Polym. Sci. A-1, 5, 175 (1967).

[56] *H. H. G. Jellinek*, "Reactions of Polymers with Pollutant Gases" in [2].

[57] *E. Weiss*, "Testing of Ozone Resistance of Polymers" in *J. V. Schmitz* (ed.), "Testing of Polymers", Interscience, New York (1966).

[58] ASTM D 470-64T,   ASTM D 518-61,   ASTM D 574-62,   ASTM D 1149-64,   ASTM D 1352-60, ASTM D 1373-64T; B.S. 2899 (1958) and (1961); DIN 53509.

[59] (a) *A. Agster*, Melliand Textilber. 47, 1279 (1966);
     (b) *E. Naujoks* and *P. Ney*, Melliand Textilber. 57, 401 (1976).

[60] *J. F. Rabek*, "Oxidative Degradation of Polymers", in *C. H. Bamford* and *C. F. H. Tipper* (eds.), "Comprehensive Chemical Kinetics", Vol. 14, Elsevier, Amsterdam (1975).

[61] *Y. Kamiya* and *E. Niki*, "Oxidative Degradation", Ref. [2], p. 79.

[62] see e.g. *K. Wiberg*, "Oxidation in Organic Chemistry", Parts A und B, Academic Press, New York (1965).

[63] *L. Reich* and *S. S. Stivala*, "Autoxidation of Hydrocarbons and Polyolefins", Dekker, New York (1969).

[64] *H. Herlinger*, *M. Sodnik*, *W. Aichele* and *E. Schollmeyer*, Melliand Textilber. 9, 743 (1977)·

[65] *D. N. Rampley* and *J. A. Hasnip*, J. Oil. Col. Chem. Ass. 59, 356 (1976).

[66] *T. A. Girard*, *M. Beispiel* and *C. E. Brieker*, J. Am. Oil Chem. Soc. 42, 828 (1965).

[67] *E. R. Mueller*, Ind. Eng. Chem. 46, 564 (1954).

[68] *D. L. Allara* and *C. W. White*, J. Am. Chem. Soc. Div. Polym. Chem., Polym. Prepr. 18, 482 (1977).

[69] *Z. Osawa* and *T. Saito*, "The Effect of Transition Metal Compounds on the Thermal Oxidative Degradation of Polypropylene in Solution", in [4], p. 159.

[70] *V. M. Yurev*, *A. N. Pravednikov* and *S. S. Medvedev*, J. Polym. Sci. 55, 353 (1961).

[71] *F. H. Winslow*, "Physical Factors in Polymer Degradation and Stabilization" in [3], p. 11.

[72] *V. S. Pudov*, *B. A. Gromov*, *M. B. Neiman* and *E. F. Sklyarova*, Neftekhimiya 3, 543, 743, and 750 (1963).

[73] *E. N. Matveeva*, *S. S. Khnikis*, *A. I. Tsevtkova* and *V. A. Balandina*, Plastomassy v. Masshimostor, Mosk. Dom Nauk i Propagandy, Vol. 2 (1963).

[74] *J. C. W. Chien*, J. Polym. Sci. A-1, 6, 375 (1968); *J. C. W. Chien* and *H. Jabloner*, J. Polym. Sci. A-1, 6, 393 (1968).

[75] *E. M. Bevilaqua* and *P. M. Norling*, Science 147, 289 (1965).

[76] *J. H. Adams*, J. Polym. Sci. A-1, 8, 1077 (1970).

[77] (a) *J. C. W. Chien* and *C. R. Boss*, J. Polym. Sci. A-1, 5, 3091 (1967);
     (b) *J. C. W. Chien* and *D. S. T. Wang*, Macromolecules 8, 920 (1975).

[78] *J. A. Howard* and *J. E. Bennet*, Can. J. Chem. 50, 2374 (1972).

[79] *Y. T. Denisov*, Vysokomol. soyed. A 21, 527 (1979), english translation: Polym. Sci. U.S.S.R. 21, 577 (1979).

[80] *J. L. Bolland* and *H. Hughes*, J. Chem. Soc. 492 (1949).

[81] *E. M. Bevilacqua*, J. Am. Chem. Soc. 77, 5396 (1955); 79, 2915 (1957) and 80, 5346 (1958).

[82] *E. M. Bevilaqua*, *E. S. English* and *E. E. Philipp*, J. Org. Chem. 25, 1276 (1960).

[83] *J. L. Morand*, Rubber Chem. Technol. 47, 1094 (1974); and 50, 773 (1977).

[84] *J. C. W. Chien* and *J. K. Y. Kiang*, Macromolecules 12, 1088 (1979).

[85] *J. R. Shelton*, *R. L. Pecsok* and *J. L. Koenig*, "Fourier Transform IR Studies on the Uninhibited Autoxidation of Elastomers" in [3], p. 75.

[86] *H. Staudinger* und *E. O. Leupold*, Chem. Ber. 63, 730 (1930); J. prakt. Chem. 161, 20 (1942).

[87] *G. V. Schulz*, Chem. Ber. 80, 335 (1947).

[88] *R. L. Whistler* and *J. N. DeMiller*, Adv. Carbohydrate Chem. 13, 289 (1958).

[89] *B. Vollmert*, "Grundriß der Makromolekularen Chemie", Springer (1962); (English Translation: *B. Vollmert*, "Polymer Chemistry", Springer, New York (1973)); Makromol. Chem. 5, 110 (1950).

[90] *A. Meller*, Holzforschung 14, 78 and 129 (1960).

[91] *D. B. Mutton*, Pulp Paper Mag. Can. 65, T 44 (1964).

[92] *A. R. Procter* and *R. H. Wiekenkamp*, J. Polym. Sci. Symposium 28, 1 (1969).

[93] *Yuan-Zong Lai* and *K. V. Sarkanen*, J. Polym. Sci. Symposium 28, 15 (1969).

[94] *A. Palma*, *S. Jovanović* and *G. V. Schulz*, Makromol. Chem. 179, 395 (1978); J. Polym. Sci., Symposium 42, 1499 (1973).

[95] *I. Ziderman* and *J. Bel-Ayche*, J. Appl. Polym. Sci. 22, 711 and 1151 (1978).
[96] *M. H. Johansson* and *O. Samuelson*, J. Appl. Polym. Sci. 22, 615 (1978).
[97] *G. Delzenne* and *G. Smets*, Makromol. Chem. 23, 16 (1957).
[98] *J. Mejzlik*, Makromol. Chem. 59, 184 (1963).
[99] *E. N. Kumpanenko, A. I. Varshavskaya, L. V. Karmilowa* and *N. S. Enikolopyan*, J. Polym. Sci. A-1, 2375 (1970).
[100] (a) *R. P. Simonds* and *E. J. Goethals*, Makromol. Chem. 179, 1689 and 1851 (1978);
      (b) private communication;
      (c) Prepr. Eur. Disc. Meet. Polym. Sci. Strasbourg (1978), page 106.
[101] *M. L. Hallensleben* and *K. Möller*, Polym. Bull. 1, 199 (1978).
[102] *J. D. Hoffman* and *G. T. Davis*, J. Res. Nat. Bur. Stand. 79 A, 613 (1975).
[103] *H. G. Zachmann*, Z. Naturforsch. 19 a, 1937 (1964).
[104] *A. Keller, E. Martuscelli, D. J. Priest* and *Y. J. Udagawa*, J. Polym. Sci. A-2, 9, 1807 (1971).
[105] *P. J. Flory*, J. Am. Chem. Soc. 84, 2857 (1962).
[106] *R. P. Palmer* and *A. J. Cobbold*, Makromol. Chem. 74, 174 (1964).
[107] *D. J. Priest*, J. Polym. Sci. A-2, 9, 1777 (1971).
[108] *M. H. Johansson* and *O. Samuelson*, J. Appl. Polym. Sci. 22, 615 (1978).
[109] *S. Ševčik, M. Kubin* and *J. Stamberg*, Proc. 5th IUPAC Conference on Modified Polymers, Bratislava (1979), Vol. I, p. 241.
[110] *N. Okui* and *J. H. Magill*, Polymer 17, 1086 (1976).
[111] *B. G. Rånby* and *E. Ribi*, Experientia 6, 12 (1950).
[112] *A. Keller* and *S. Sawada*, Makromol. Chem. 74, 190 (1964).
[113] *A. Peterlin* and *G. Meinel*, J. Polym. Sci. B 3, 1059 (1965).
[114] *C. W. Hock*, J. Polym. Sci. B 3, 573 (1965); A-2, 4, 227 (1966).
[115] *R. S. J. Manley*, J. Polym. Sci. Polym. Phys. Ed. 12, 1347 (1974).
[116] *J. L. Koenig* and *M. Hannon*, J. Macromol. Sci. B 1, 119 (1967).
[117] *T. Matsumoto, N. Ikegami, K. Ehara, T. Kawai* and *H. Maeda*, Kogyo Kagaku Zasshi 73, 2441 (1970).
[118] *K. H. Illers*, Makromol. Chem. 118, 88 (1968).
[119] *J. J. Breedon* and *P. H. Geil*, J. Res. Nat. Bur. Stand. 79 A, 609 (1975).
[120] *A. M. Basedow, K. H. Ebert* and *H. J. Ederer*, Macromolecules 11, 774 (1978).
[121] *I. Sakurada, Y. Sakagushi* and *M. Kagau*, Kobunshi Kagaku 17, 87 (1960).
[122] *I. Sakurada*, Pure Appl. Chem. 16, 263 (1968).
[123] *R. Brdička*, "Physical Chemistry Fundamentals", Natural Science Publishing House Prague (1952), p. 541.
[124] *H. H. G. Jellinek*, "Reactions of Polymers with Pollutant Gases" in [2], p. 431.
[125] *H. H. G. Jellinek, F. Flajsman* and *F. J. Kryman*, J. Appl. Polym. Sci. 13, 107 and 2504 (1969).
[126] *H. H. G. Jellinek* and *T. J. Y. Wang*, J. Polym. Sci., Polym. Chem. Ed. 11, 3227 (1973).
[127] *T. Ogihara*, Bull. Chem. Soc. Jap. 36, 58 (1963); *T. Ogihara, T. Tsuchiya* and *K. Kuratani*, Bull. Chem. Soc. Jap. 38, 978 (1965).
[128] *H. Kachi* and *H. H. G. Jellinek*, J. Polym. Sci., Polym. Chem. Ed. 17, 2031 (1979).
[129] *I. Mellan*, "Corrosion Resistant Materials Handbook", Noyes Development Corp., Park Ridge (USA) (1966).
[130] *H. Saechtling*, "Kunststoff-Taschenbuch", 21th edition, Hanser, München (1979).
[131] *C.-M. v. Meysenburg*, Kunststoffe — Fortschrittsberichte Vol. 3, Part 2, Hanser, München (1976); see also [1], p. 373 and [28].
[132] *S. Yamashita, N. Kawabata, S. Sagan* and *K. Hayashi*, J. Appl. Polym. Sci. 21, 2201 (1977).
[133] *N. Kawabata, S. Yamashita* and *Y. Furukawa*, Bull. Chem. Soc. Jap. 51, 625 (1978).
[134] Verband Kunststofferzeugende Industrie, Frankfurt/M. (ed.), "Hydrolysis of Scrap Plastics", Final Report (in German) by *E. Grigat, G. Niederdellmann, H. Hetzel* and *A. Bergmann-Franke*, Leverkusen/Dormagen (1979).

Index